STUDENT UNIT GUIDE

AQA | A2 | UNIT 5

Biology
Control in Cells and in Organisms

Steve Potter

A2 Biology

Philip Allan Updates, an imprint of Hodder Education, an Hachette UK company, Market Place, Deddington, Oxfordshire OX15 0SE

Orders
Bookpoint Ltd, 130 Milton Park, Abingdon, Oxfordshire OX14 4SB
tel: 01235 827720
fax: 01235 400454
e-mail: uk.orders@bookpoint.co.uk
Lines are open 9.00 a.m.–5.00 p.m., Monday to Saturday, with a 24-hour message answering service. You can also order through the Philip Allan Updates website: www.philipallan.co.uk

© Philip Allan Updates 2010

ISBN 978-0-340-94953-5

First printed 2010
Impression number 5 4 3 2 1
Year 2014 2013 2012 2011 2010

All rights reserved; no part of this publication may be reproduced, stored in a retrieval system, or transmitted, in any other form or by any means, electronic, mechanical, photocopying, recording or otherwise without either the prior written permission of Philip Allan Updates or a licence permitting restricted copying in the United Kingdom issued by the Copyright Licensing Agency Ltd, Saffron House, 6–10 Kirby Street, London EC1N 8TS.

This guide has been written specifically to support students preparing for the AQA A2 Biology Unit 5 examination. The content has been neither approved nor endorsed by AQA and remains the sole responsibility of the author.

Typeset by Philip Allan Updates
Printed by MPG Books, Bodmin

Hachette UK's policy is to use papers that are natural, renewable and recyclable products and made from wood grown in sustainable forests. The logging and manufacturing processes are expected to conform to the environmental regulations of the country of origin.

Contents

Introduction
About this guide 4
Preparing for the Unit 5 examination 5
Approaching the Unit 5 examination 6

■ ■ ■

Content Guidance
About this section 10
How organisms detect and respond to changes in their environment 11
How organisms coordinate their responses to stimuli 21
Skeletal muscle: a key effector in mammals 43
Homeostasis and feedback systems 48
The genetic code, protein synthesis and gene mutation 58
The control of gene action 69
Gene cloning technology and its applications 73

■ ■ ■

Questions and Answers
About this section 86
Q1 Hormone action 87
Q2 Action potentials 89
Q3 Protein synthesis 91
Q4 The polymerase chain reaction 93
Q5 Gene cloning technology and its applications 95
Q6 The control of gene action 97
Q7 Auxins 99
Q8 Skeletal muscle 101
Q9 Receptors and transmission of information through the nervous system 103
Q10 Homeostasis 108

Introduction

About this guide

This guide is written to help you to prepare for the Unit 5 examination of the new AQA biology specification. The Unit 5 examination examines the content of **Unit 5: Control in cells and in organisms**, and forms part of the A2 assessment.

This **Introduction** provides guidance on revision, together with advice on approaching the unit examination.

The **Content Guidance** section gives a point-by-point description of all the facts you need to know and concepts you need to understand for Unit 5. In each topic, the concepts are presented first. It is a good idea to get your mind around these key ideas before you try to learn all the associated facts.

The **Questions and Answers** section shows you the sort of questions you can expect in the unit examination. It would be impossible to give examples of every kind of question in one book, but these should give you a flavour of what to expect. Each question has been attempted by two candidates, Candidate A and Candidate B. Their answers, along with the examiner's comments, should help you to see what you need to do to score a good mark — and how you can easily *not* score a mark even though you may understand the biology.

What can I assume about the guide?

You can assume that:
- the basic facts you need to know and understand are stated explicitly
- the major concepts you need to understand are explained clearly
- the questions at the end of the guide are similar in style to those that will appear in the end-of-unit examination.
- the answers supplied are the answers of A2 students
- the standard of the marking is broadly equivalent to the standard that will be applied to your answers

What can I *not* assume about the guide?

You *must not* assume that:
- the diagrams used will be the same as those used in the end-of-unit examination (they may be more or less detailed, seen from a different angle etc.)
- the way in which the concepts are explained is the *only* way in which they can be presented in an examination (concepts are often presented in an unfamiliar situation)
- the range of question types presented is exhaustive (examiners are always thinking of new ways to test a topic)

How Science Works

This is a new component in the biology specifications of all examination boards. The aim is to help you to understand the *process* of scientific work. You will not find any specific section devoted to 'How Science Works' (HSW) in this guide, but the main aspects of HSW are described below.

- Scientists use pre-existing knowledge and understanding/theories/models to suggest explanations for phenomena.
- They design, carry out, analyse and evaluate scientific investigations to test new explanations.
- They share their findings with other scientists so that they may be validated, or not as the case may be.

As a consequence of the work of scientists, there may be implications for society as a whole. You are expected to appreciate and make informed (not emotional) comment on such aspects as:

- the ethical implications of the way in which research is carried out
- the way in which society uses science to help in decision making

Some of the questions in the Questions and Answers section address HSW.

So how should I use this guide?

The guide lends itself to a number of uses throughout your course — it is not *just* a revision aid. You can use it:

- to check that your notes cover the material required by the specification
- to identify strengths and weaknesses
- as a reference for homework and internal tests
- during your revision to prepare 'bite-sized' chunks of related material, rather than being faced with a file full of notes

The Questions and Answers section can be used to:

- identify the terms used by examiners in questions and what they expect of you
- familiarise yourself with the style of questions you can expect
- identify the ways in which marks are lost as well as how they are gained

Preparing for the Unit 5 examination

Preparation for examinations is a personal thing. Different people prepare, equally successfully, in very different ways. The key is being honest about what actually *works* for *you*.

Whatever your style, you must have a plan. Sitting down the night before the examination with a file full of notes and a textbook does not constitute a revision plan — it is just desperation, and you must not expect a great deal from it. Whatever your personal style, there are a number of things you *must* do and a number of other things you *could* do.

Things you *must* do
- Leave yourself enough time to cover all the material.
- Make sure that you actually *have* all the material to hand (use this book as a basis).
- Identify weaknesses early in your preparation so that you have time to do something about them.
- Familiarise yourself with the terminology used in examination questions (see below).

Things you *could* do to help you learn

Psychologists have shown that you learn facts and ideas better if you are *active* in your learning. Just reading your notes over and over again is not a very good way of revising. Instead, you could:
- write a summary of your notes which includes all the key points
- write key points on postcards (carry them round with you for quick revision during a coffee break)
- discuss a topic with a friend who is studying the same course
- try to explain a topic to someone not on the course
- practise examination questions on the topic

All these techniques make you *think* about the material. The more you *process* the information as you revise, the more effective your revision will be.

Approaching the Unit 5 examination

Terms used in examination questions

You will be asked precise questions in the examinations, so you can save a lot of valuable time and ensure that you score as many marks as possible by knowing what is expected. Terms most commonly used are explained below.
- **Describe** — this means exactly what it says — 'tell me about...' — and you should not need to explain why.
- **Explain** — give biological reasons for *why* or *how* something is happening.
- **Complete** — finish off a diagram, graph, flow chart or table.
- **Draw/plot** — construct some type of graph. For this, make sure that:
  - you choose a scale that makes good use of the graph paper (if a scale is not given) and does not leave all the plots tucked away in one corner
  - plot an appropriate type of graph — if both variables are continuous variables, then a line graph is usually the most appropriate; if one is a discrete variable, then a bar chart is appropriate
  - plot carefully using a sharp pencil and draw lines accurately
- **From the**... — use only information in the diagram/graph/photograph or other forms of data.

- **Name** — give the name of a structure/molecule/organism etc.
- **Suggest** — i.e. 'give a plausible biological explanation for'; this term is often used when testing understanding of concepts in an unfamiliar situation.
- **Compare** — give similarities *and* differences between...
- **Calculate** — add, subtract, multiply, divide (do some kind of sum) and show how you got your answer — *always* show your working.

The examination

When you finally open the test paper, it can be quite a stressful moment. You may not recognise the diagram or graph used in question 1. It can be quite demoralising to attempt a question at the start of an examination if you are not feeling very confident about it. The following advice should help you achieve a good result.
- Do *not* begin to write as soon as you open the paper.
- Do *not* necessarily answer question 1 first, just because it is printed first (the examiner did not sequence the questions with your particular favourites in mind).
- Scan *all* the questions before you begin to answer any.
- Identify those questions about which you feel most confident.
- Answer *first* those questions about which you feel most confident regardless of the order in the paper.
- *Read the question carefully* — if you are asked to explain, then explain, don't just describe.
- Take notice of the mark allocation. Don't supply the examiner with all your knowledge of osmosis if there is only 1 mark allocated (similarly, you have to come up with four ideas if 4 marks are allocated).
- Try to stick to the point in your answer (it is easy to stray into related areas that will not score marks and that will use up valuable time).
- Take particular care with:
  - drawings — you will not be asked to produce complex diagrams, but those you do produce must resemble the subject
  - labelling — label lines *must touch* the part you are required to identify; if they stop short or pass through the part, you will lose marks
  - graphs — draw *small* points if you are asked to plot a graph and join the plots with ruled lines or, if specifically asked for, a line or smooth curve of best fit through all the plots
- Try to answer *all* the questions.

Content Guidance

A2 Biology

This section is a guide to the content of **Unit 5: Control in cells and in organisms**. The main areas of this module are:
- how organisms detect and respond to changes in their environment
- how organisms coordinate their responses to stimuli
- skeletal muscle — a key effector in mammals
- homeostasis and feedback systems
- the genetic code, protein synthesis and gene mutation
- the control of gene action
- gene cloning technology and its applications

Key facts you must know and understand

These are exactly what you might think: a summary of all the basic knowledge that you must be able to recall and show that you understand. The knowledge has been broken down into a number of small facts that you must learn. This means that the list of 'Key facts' for some topics is quite long. However, this approach makes quite clear *everything* you need to know about the topic.

Key concepts you must understand

These are a little different. Whereas you can learn facts, these are ideas or concepts that often form the basis of models that we use to explain aspects of biology. You can know the actual words that describe a concept like osmosis, or the resolving power of a microscope, but you will not be able to use this information unless you really understand what is going on. Once you genuinely understand a concept, you will probably not have to learn it again.

What the examiners could ask you to do

This part tries to give you an insight into the minds of the examiners who will set and mark your examination papers. They may ask you to recall any of the basic knowledge or explain any of the key concepts; but they may well do more than that. Examiners think up questions where the concepts you understand are in a different setting or context from the one(s) you are familiar with. This could include the evaluation of data under the 'How Science Works' requirement, set in the context of this particular topic.

Bear in mind that examiners will often set individual questions that involve knowledge and understanding of more than one section. The sample questions in the Questions and Answers section of this book will help you to practise drawing together material from different areas of the specification.

How organisms detect and respond to changes in their environment

The survival value of behaviour patterns

Key concepts you must understand

Organisms respond to changes in their external environment and to changes in their internal environment.

When an organism responds to a **stimulus** (a change in the external or internal environment of an organism) there are a number of processes that are common to all, no matter how simple or complex. There is always:
- a **receptor** — a structure that detects the stimulus
- an **effector** — a structure, such as a muscle, that produces the response
- some kind of **linking system** or **coordinating system** — this receives information from the receptor and passes information to the effector
- a **response** — the action that results from the stimulus

This is shown in Figure 1.

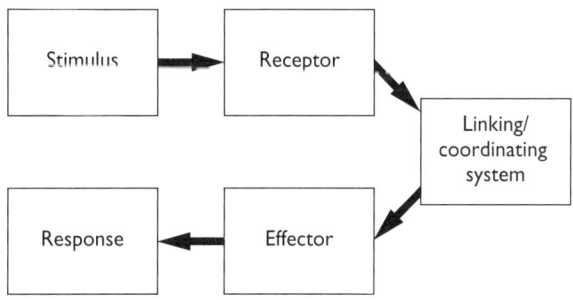

Figure 1 The processes involved in responding to a stimulus

Many examples of behaviour have evolved to increase the chances of survival of an organism. These include the following:
- Plant shoots (stems) grow towards the region of most intense light; this exposes the leaves to the maximum amount of light and increases the rate of photosynthesis.
- Woodlice move more quickly in the light than in the shade; this increases their chances of moving out of the light and into the shade (where it is also likely to be more humid, so the animals will not dehydrate as quickly).

- Reflex actions in mammals are often protective — for example:
  - withdrawal reflexes move parts of the body away from potentially damaging stimuli, such as heat or sharp objects
  - when a bright light is shone into the eye, the iris reflex prevents too much light from entering the eye and damaging the retina (see Figure 2)

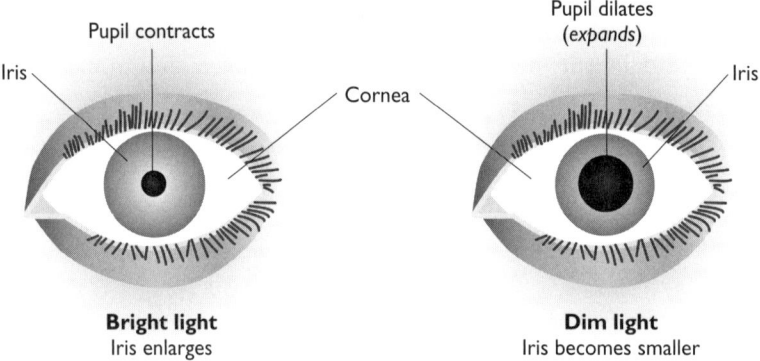

Figure 2 The iris reflex

Other reflex actions are corrective rather than protective and include:
- reducing the heart rate if blood pressure becomes too high
- increasing the breathing rate if the partial pressure of oxygen in the blood becomes too low or the partial pressure of carbon dioxide becomes too high

Key facts you must know and understand

Responses of plants to external stimuli often involve growth. These responses are called **tropisms**. They are usually responses to differences in the intensity of a stimulus from different directions. A growth response *towards* a stimulus is a positive tropism; a growth response *away* from a stimulus is a negative tropism. For example:
- plant shoots grow towards the most intense source of light (positive **phototropism**)
- plant shoots grow away from gravity (negative **gravitropism**)

In simple animals, there are two main types of response to stimuli:
- **Taxis** (plural **taxes**), in which the animal moves along a gradient of intensity of a stimulus either towards the greatest intensity of the stimulus (a positive taxis) or away from the greatest intensity (a negative taxis); there is a directional response to a directional stimulus. Maggots move away from light; many insects are attracted by, and move towards, pheromones (chemicals similar to hormones).
- **Kinesis** (plural **kineses**), in which a change in the intensity of the stimulus brings about a change in the rate of movement, *not* a change in the direction of movement. Woodlice move around more in dry conditions and less in moist conditions. Therefore, they are more likely to move out of dry conditions and remain in moist conditions, where they are less likely to dehydrate.

Both of these are examples of instinctive, non-intelligent behaviour.

Variation in results

Investigating the behaviour of animals produces results with more variability than the results from an investigation into, say, enzyme activity. There are two main reasons for this:

- Only 20 or so woodlice are used, rather than billions of molecules of enzyme; one woodlouse that behaves differently has a bigger effect on the results than one enzyme molecule that behaves differently.
- There are genetic differences between woodlice and some of these influence behaviour.

The preference of woodlice for a dark or a light environment can be investigated using a simple choice chamber.

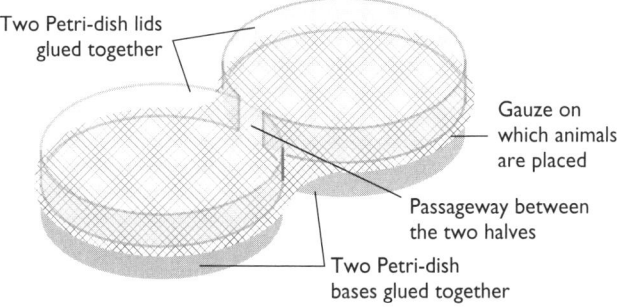

Figure 3 A simple choice chamber

Ten woodlice are placed in each half of the choice chamber. One half of the chamber is covered with black paper and the apparatus is left for 10 minutes. At the end of this time, the number of woodlice in each environment (dark and light) is noted. To help overcome the small sample size, the experiment is repeated several times and mean numbers for each environment are calculated.

The results can then be expressed as a bar chart. To give some idea of the variability of the results, the standard error (SE) of each mean is calculated.

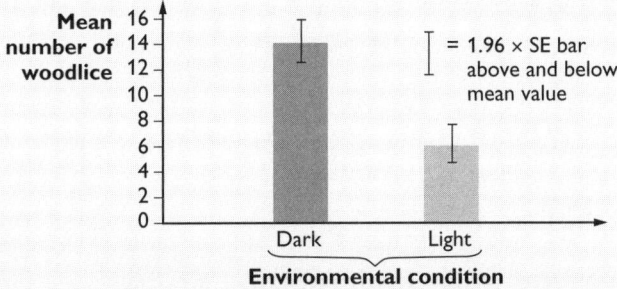

Figure 4 Bar chart of choice-chamber results

A2 Biology

Error bars with the value of 1.96 × SE are shown above and below the mean value.

If the error bars for the two values overlap, then we must conclude that there is a greater than 5% probability that the differences are due to chance and have no underlying cause. If the error bars do not overlap, then we can conclude that there is less than 5% probability that the differences are due to chance and that there *is* an underlying cause for the differences.

> **Tip** Notice the way in which the terms 'probability' and 'chance' are used. Don't confuse them — they are not the same.

Reflex actions in mammals are unlearned (instinctive), automatic responses to a stimulus. They are controlled by nerves. There are two main types of reflex action:
- **a somatic reflex** is a response to an external stimulus — for example, the iris reflex and the withdrawal reflexes
- **an autonomic reflex** is a response to an internal stimulus — for example, the reflexes controlling heart rate and breathing rate

Receptors

Key concepts you must understand

In humans, receptors are specialised sense cells that act as **energy transducers**. That is, they use a specific type of energy to produce a **generator potential**. This is a small change in voltage in a membrane in the sense cell. This may then initiate an **action potential** in a nerve cell. An action potential is a small voltage change in the plasma membrane of a nerve cell. Once an action potential has been initiated, it sweeps along the nerve cell. This passage of an action potential along a nerve cell is a **nerve impulse**.

Nerve impulses are 'all-or-nothing' events. There is a certain level of stimulation needed to initiate a nerve impulse. This is called the **threshold** level. If the generator potential produced in the sense cell does not stimulate a nerve cell above the threshold value, there is no nerve impulse.

Different receptors transduce different types of energy.

Detecting changes in the internal environment

Key concepts you must understand

Nearly all the reactions that take place in our bodies are controlled by enzymes. It is important that the enzymes work with maximum efficiency. To achieve this, they must be in an environment that is maintained at, or very near to, their optimum temperature and pH. Other factors must also be maintained at appropriate levels.

Maintaining a constant internal environment is called **homeostasis** and often involves **negative feedback** systems. For more details on both of these, see pages 48–58.

To be able to maintain a constant internal environment, it is essential that any changes are detected. The sensors are specific; they detect only changes in one particular factor.

Key facts you must know and understand

Humans have sensors that detect changes in many factors, including:
- core body temperature
- plasma glucose concentration
- blood pressure
- the partial pressure of carbon dioxide in the plasma

Changes in the partial pressure of carbon dioxide in the plasma and changes in blood pressure are important in the regulation of heart rate. Sensors that monitor both are found in the aortic arch. **Baroreceptors** monitor changes in blood pressure; **chemoreceptors** monitor changes in the partial pressure of carbon dioxide in the plasma. The location of these sensors is shown in Figure 5.

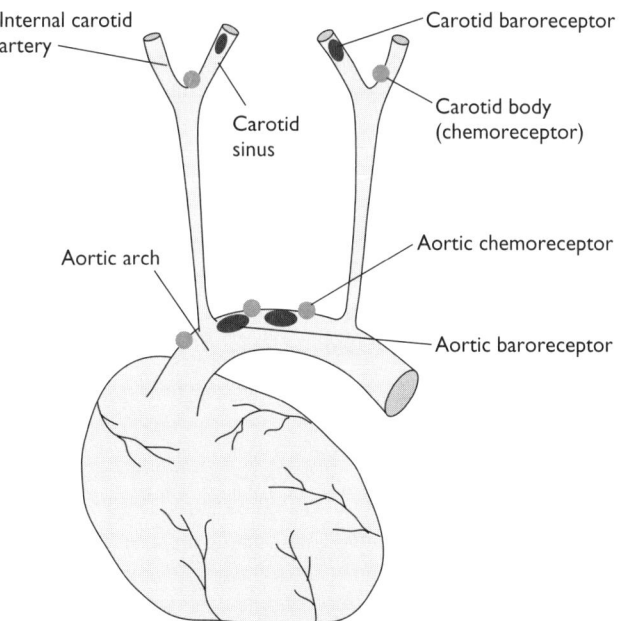

Figure 5 The location of baroreceptors and chemoreceptors close to the heart

The flow charts in Figure 6 show how changes in the partial pressure of carbon dioxide (pCO_2) and in arterial pressure influence the heart rate.

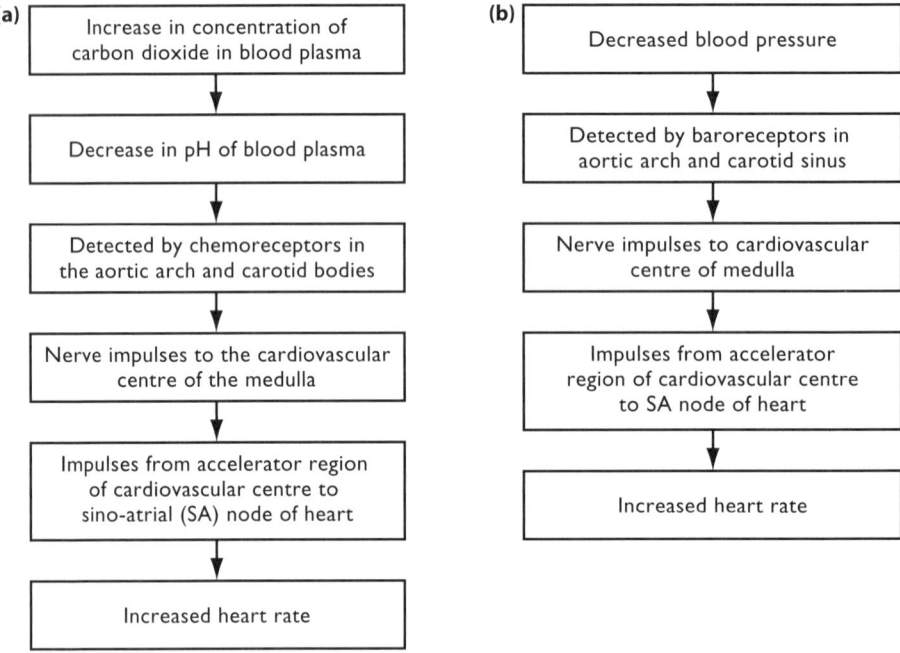

Figure 6 (a) The effect of an increase in pCO_2 on heart rate
(b) The effect of a decrease in arterial blood pressure on heart rate

Detecting changes in the external environment: pressure on the skin

Key concepts you must understand

Pressure is detected by several different receptors in the skin. One of these is the **Pacinian corpuscle** (Figure 7).

The membrane that acts as the energy transducer is the plasma membrane of a sensory nerve cell that is surrounded by many layers of lamellae separated by a thick gel. If sufficient pressure is transmitted to the membrane then pressure-sensitive sodium ion channels open and sodium ions enter the nerve cell. Because sodium ions are positively charged, the voltage inside and outside the membrane changes — a generator potential is produced (see Figure 8).

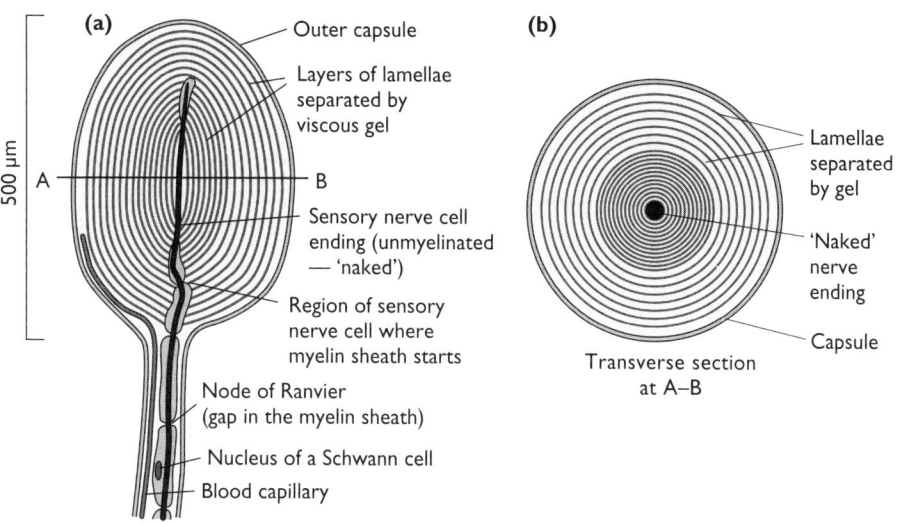

*Figure 7 (a) A Pacinian corpuscle seen in longitudinal section
(b) A Pacinian corpuscle seen in transverse section*

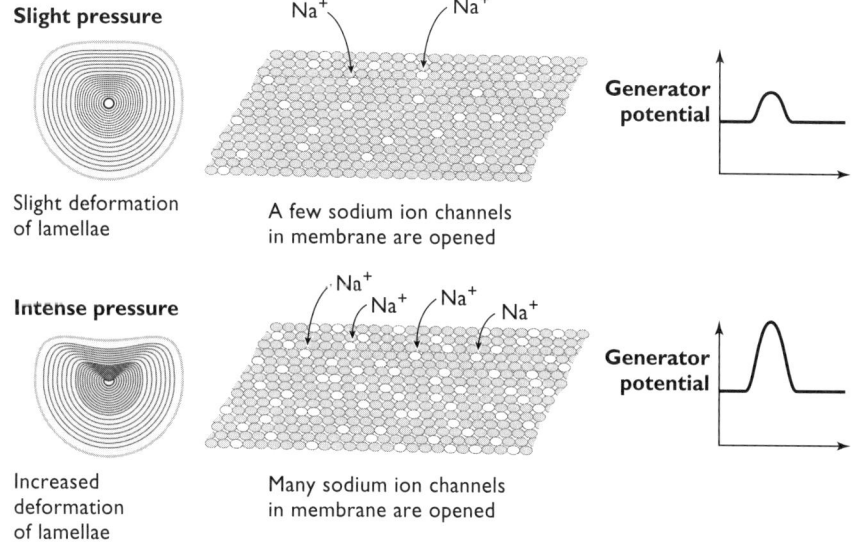

Figure 8 The effect of pressure on the generator potential produced by a Pacinian corpuscle

Greater pressure means that more sodium ion channels open and a bigger generator potential results.

A2 Biology

The lamellar structure of the corpuscle is important in ensuring that only quite firm pressure stimulates the naked nerve ending. Light pressure deforms the lamellae slightly, but most of this pressure is then absorbed by the gel and not transmitted to the nerve ending at the centre of the corpuscle.

Tip Do *not* say that the gel insulates the nerve ending. Insulation can apply to heat and electricity, but not to pressure.

Detecting changes in the external environment: light intensity and quality

Key concepts you must understand

Sense cells in the **retina** of the eye contain photosensitive pigments that degrade when struck by light. **Rod cells** contain **rhodopsin** and **cone cells** contain **iodopsin**. The change in the pigment produces a change in the voltage of a membrane in the rod or cone and results in a generator potential (see Figure 9).

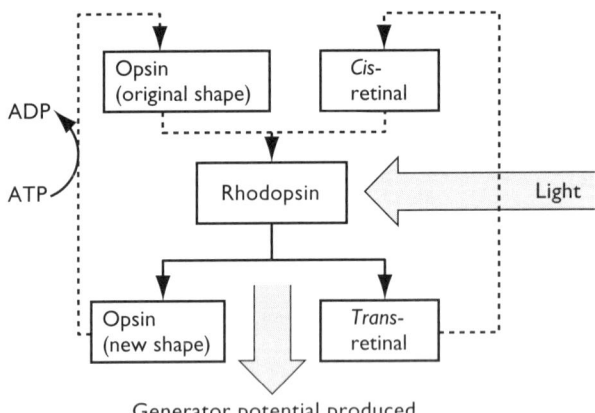

Figure 9 How rhodopsin produces a generator potential when struck by light

Rods and cones differ in their:
- **acuity** — the amount of detail in which something is perceived
- **sensitivity** — the intensity of light that produces a sufficiently large generator potential to result in a nerve impulse (action potential)

Both of these differences result from the way in which the rods and cones synapse with (are linked to) cells called **bipolar cells**. Each bipolar cell, in turn, synapses with one sensory nerve cell. This is illustrated in Figure 10.

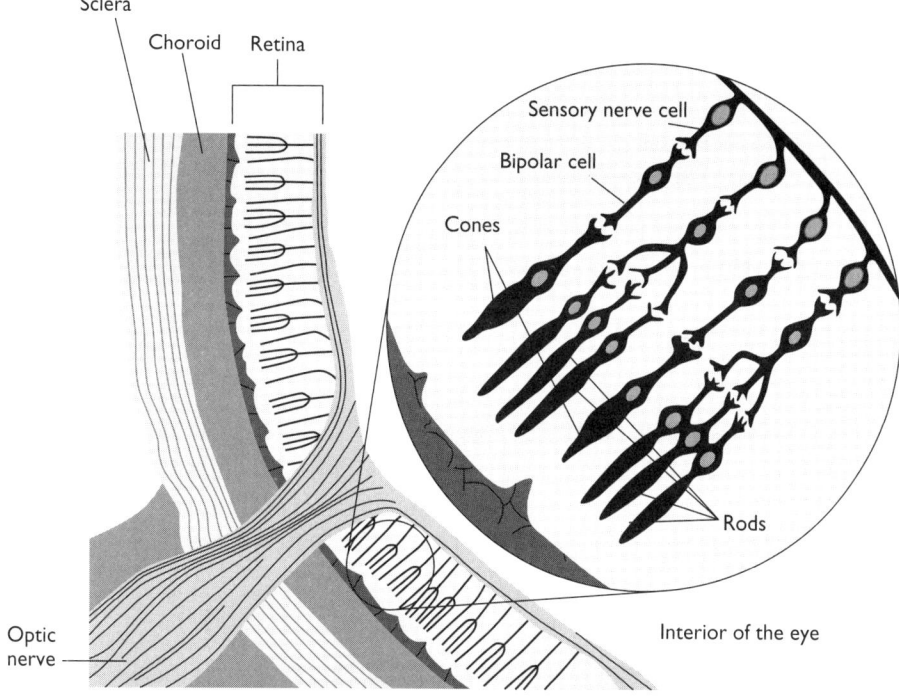

Figure 10 The way in which rods and cones synapse with bipolar cells in the retina

Each cone cell synapses with one bipolar cell, which then initiates a nerve impulse in one sensory nerve cell. There are two consequences to this:
- The single cone cell must stimulate the bipolar cell sufficiently to exceed the threshold. This is unlikely to happen in dim light and so cones have poor sensitivity.
- The area of the visual field detected by the single cone is interpreted by the brain as a single point (think of a pixel) in the image it produces. This gives the cones good acuity.

Several rod cells synapse with the same bipolar cell. This is called **retinal convergence** and there are two consequences to this:
- The generator potentials from several rod cells combine or 'add up' to stimulate the bipolar cell. This is called **summation** — it makes it more likely that the threshold will be exceeded in dim light and gives the rods good sensitivity.
- Light striking several rod cells results in a single impulse along just one sensory nerve cell to the brain. The brain cannot distinguish between the parts of the image produced by the different rods and interprets them as one point. Therefore, rods have poor acuity.

This is summarised in the table below.

Property	Cones	Rods
Sensitivity	Low: light energy transduced by a single cone must produce a generator potential large enough to exceed the threshold needed for a nerve impulse. In low light intensities this is unlikely.	High: in low light intensities, generator potentials from several rods can combine and so the threshold is more likely to be exceeded and a nerve impulse initiated. This phenomenon is called summation. It is possible because several rods are linked to (or converge on) one neurone (via bipolar cells). This is called retinal convergence.
Acuity	High: each cone is connected to a single bipolar cell, so in high light intensities each cone stimulated represents a separate part of the image which can be seen in detail.	Low: several rods are connected to the same bipolar cell, so the individual parts of the image represented by each rod are merged into one — they are indistinguishable and detail is poor.

Key facts you must know and understand

Different parts of the retina have different proportions of rods and cones (see Figure 11). This affects the kind of image formed when light falls on those regions.

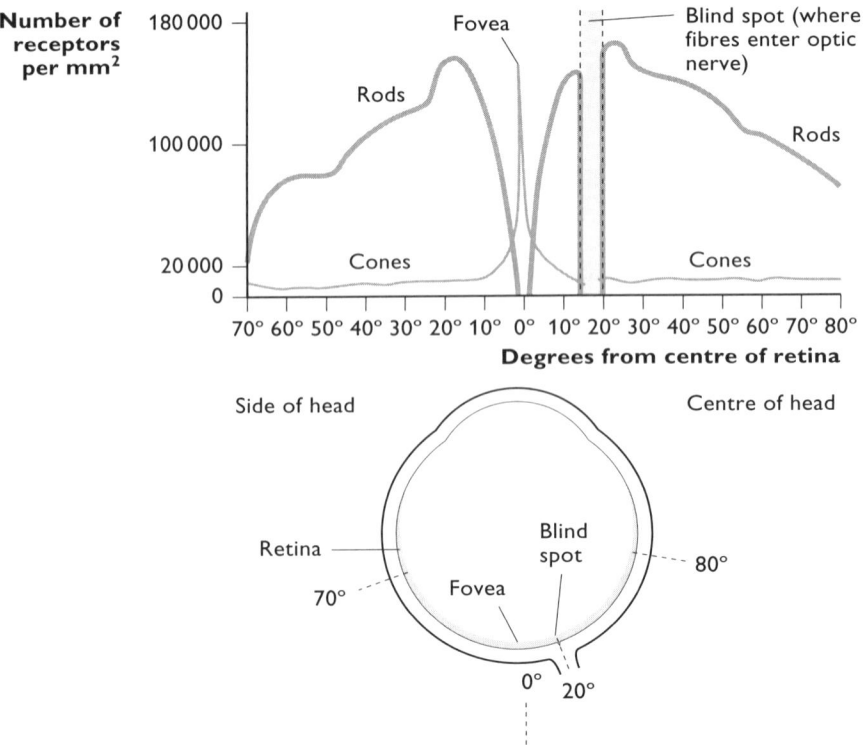

Figure 11 The distribution of rods and cones in the retina

The greatest concentration of cones is found at the **fovea** in the centre of the retina. When you look straight at an object, light from the object is focused onto the fovea. This enables the object to be seen in great detail (if the light intensity is sufficiently high).

The greatest concentration of rods is about 20° away from the fovea. At very low light intensities, looking slightly to the side of an object causes the light rays from the object to be focused on to this area of the retina. Summation by the rods allows better perception than if light fell on the fovea. So to see an object in most detail in dim light, it is best to look slightly to one side of it.

There are three different types of cone. They are sensitive to the different wavelengths of light that are broadly equivalent to the three primary colours — red, blue and green. Rods are equally sensitive to all wavelengths of light, so colour is perceived only when light falls on the cones.

What the examiners could ask you to do
- Explain any of the key concepts.
- Recall and show understanding of any of the key facts.
- Interpret data in graphs and tables showing responses to a range of stimuli.
- Interpret data concerning responses you have not studied, in terms their likely survival value to an organism.
- Use your knowledge of photosynthesis to explain the value of the phototropic and gravitropic responses of plant shoots.
- Use your knowledge of natural selection to explain how some behaviours may have become widespread in a population or species.
- Comment on the reliability and/or validity of investigations into the behaviour of organisms, bearing in mind:
    - the numbers of organisms or people used and the number of repeats carried out
    - possible flaws in the technique or in the apparatus
    - overlap of error bars in graphs of results
    - the extent to which appropriate controls were carried out

How organisms coordinate their responses to stimuli

The organisation of the nervous system

Key facts you must know and understand
The nervous system is composed of billions of cells. Most of these cells are involved in transmitting 'information' and are called **neurones**. These are organised into larger structures including nerves, the spinal cord and the brain.

The brain and the spinal cord form the **central nervous system** (**CNS**); the nerves form the **peripheral nervous system** (**PNS**). Figure 12 shows this organisation.

The nervous system is divided functionally into:
- the **somatic nervous system** (SNS), which integrates information from the special senses to produce responses in skeletal muscles
- the **autonomic nervous system** (ANS), which integrates information from receptors in internal organs and produces responses in the same or other organs or glands

The ANS is further subdivided into:
- the **sensory branch** (or division), which transmits sensory nerve impulses into the central nervous system
- the **sympathetic branch** (or division), which transmits impulses from the central nervous system to the organs, generally preparing the body for 'fight or flight' — for example by increasing cardiac output and pulmonary ventilation
- the **parasympathetic branch** (or division), which acts *antagonistically* to the sympathetic branch and prepares the body for 'rest and repair', decreasing cardiac output and pulmonary ventilation

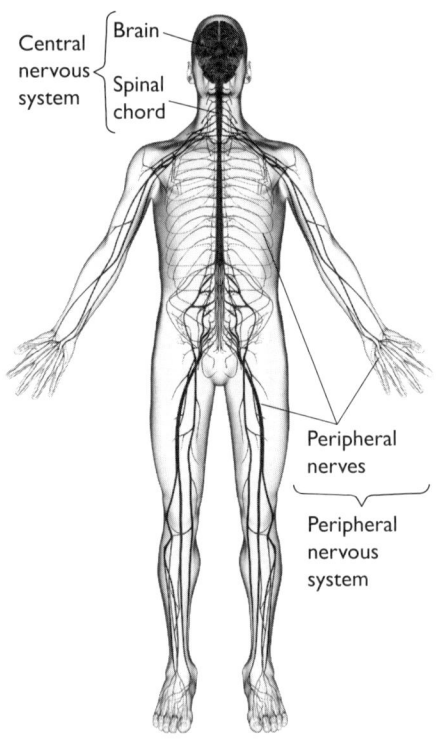

Figure 12 The structural organisation of the nervous system

The functional organisation of the nervous system is shown in Figure 13.

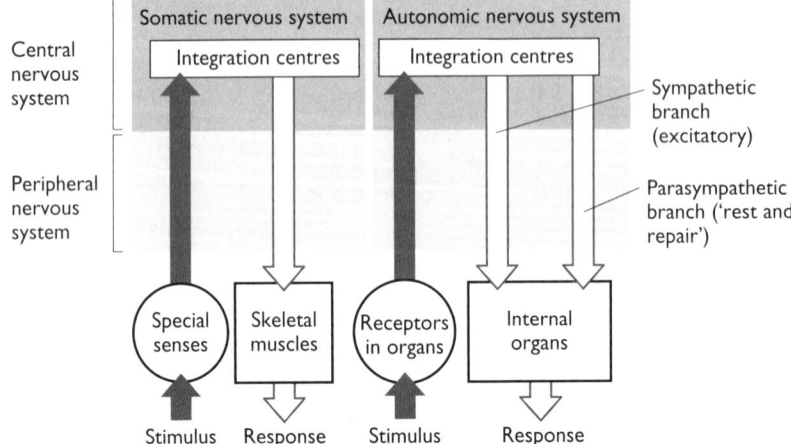

Figure 13 The functional organisation of the nervous system

The nervous system contains three main types of neurone:
- **sensory neurones** — these carry nerve impulses from sense cells into the CNS (Figure 14a)
- **motor neurones** — these carry impulses from the CNS to effectors such as muscles (Figure 14b)
- **inter-neurones** or **relay neurones** — these carry impulses from a sensory neurone to a motor neurone, often in **reflex arcs**, within the CNS

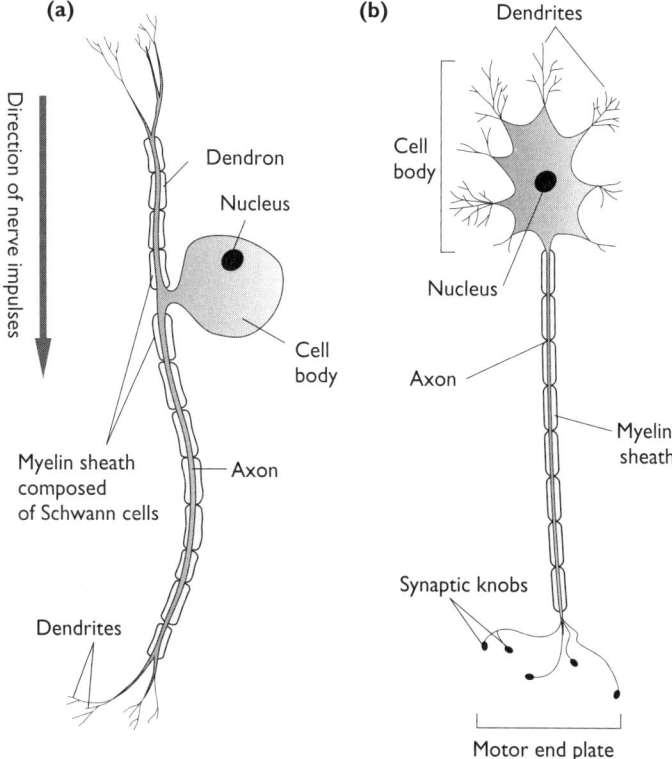

Figure 14(a) Sensory and (b) motor neurones

Impulses pass along the axons of neurones. The myelin sheath is important in ensuring that the impulses pass quickly along the axon without any deterioration.

The myelin sheath is formed as the nerve cell develops. Schwann cells, rich in myelin, wrap themselves around the developing axon. Myelin effectively insulates each axon and allows nerve impulses to pass along the axon efficiently.

In some diseases, such as multiple sclerosis, myelin is progressively lost from axons. Demyelinated axons conduct impulses much more slowly — the impulse may not reach the end of the axon. This results in a decrease in the ability to coordinate actions.

Neurones are adapted to their function of transmitting nerve impulses by having:
- long processes called axons

A2 Biology

- a specialised plasma membrane surrounding the axon that:
  - has a voltage difference (**membrane potential**) between the outside and the inside and
  - allows ions to be moved in and out of the axon, so altering the membrane potential
- many small branches called **dendrites** that synapse with other nerve cells

The resting potential

Key concepts you must understand

A nerve impulse is not like an electric current in which electrons flow through wires. What happens is that the membrane potential changes at one end of the axon, then at the next point along the axon, then the next, the next...and so on.

When a nerve impulse is not being conducted, the membrane potential of an axon is called the **resting potential** (Figure 15). In this state, the inside of the membrane is more negative than the outside by 70 millivolts (70 thousandths of a volt). This is usually written as −70 mV. Because of this difference in charge between the inside and outside of the membrane, we say that it is **polarised**.

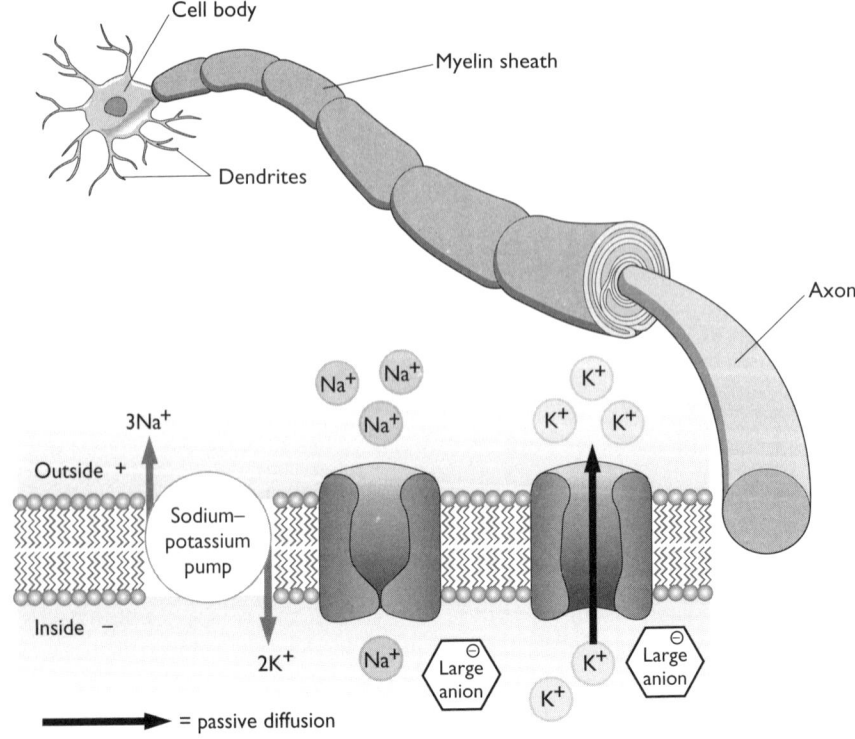

Figure 15 The resting potential

24

The resting potential is maintained by:
- large anions, for example negatively charged protein molecules inside the axon
- passive diffusion of sodium and potassium ions across the membrane through ion channels; sodium ions diffuse in more slowly than potassium ions diffuse out (both ions carry one positive charge, so the inside of the membrane loses positive charges faster than it gains them and, therefore, becomes more negative)
- active transport of sodium and potassium ions across the membrane by transport proteins; sodium ions are pumped out faster than potassium ions are pumped in (again, the inside of the membrane loses positively charged ions faster than it gains them)

Initiating an action potential

Key concepts you must understand

An action potential is initiated when either a sense cell, or another neurone, secretes a chemical called a **neurotransmitter** that crosses the **synapse** (small gap) between the secreting cell and the neurone in which the action potential will be initiated.

The neurotransmitter binds to receptors on the membrane of the neurone and influences **gated ion channels** in the axon membrane. Gated ion channels can be open or closed. They are firmly closed during the maintenance of the resting potential and play no part in this. Normally, the neurotransmitter causes the following:
- some gated sodium ion channels open
- all the gated potassium ion channels remain closed

The initiation of an action potential is shown in Figure 16.

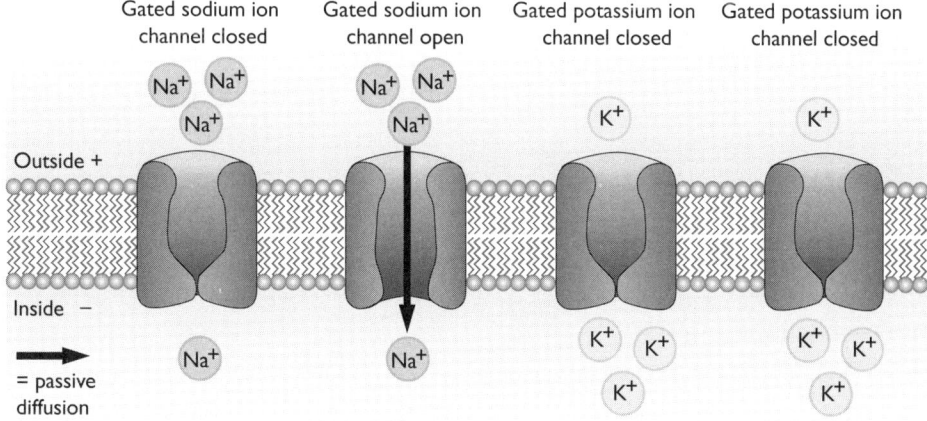

Figure 16 When an action potential is initiated, some gated sodium ion channels open, but all the gated potassium ion channels remain closed

Opening the gated sodium ion channels allows sodium ions to enter the axon. This makes the membrane potential less negative. The **threshold potential** is −55 mV. If the membrane potential reaches this level, then the other gated sodium ion channels open. Because they only open at this set voltage they are called **voltage-sensitive sodium ion channels**.

When all the voltage-sensitive sodium ion channels open, sodium ions enter much more rapidly (they 'flood in'). All the extra (positively charged) sodium ions inside the membrane raise its potential to +40 mV (with respect to the outside of the membrane).

Because of the change in the membrane potential (from −70 mV to +40 mV), we say the membrane has been **depolarised**.

Once the membrane potential reaches +40 mV, all the gated sodium ion channels close. No more sodium ions can enter the axon and the membrane potential cannot become any more positive. Because of this, action potentials always have the same value — they are all +40 mV.

The other consequence of threshold is that if the neurotransmitter only causes the membrane potential to rise to (say) −60 mV, then the voltage-sensitive sodium ion channels do not open and there is no action potential. It is an **all-or-nothing** event.

Repolarising ready for the next action potential

Key concepts you must understand

Before another action potential can be initiated, the axon membrane must return to the 'resting' condition — it must **repolarise** itself.

As soon as the action potential (+40 mV) is reached, the gated sodium ion channels close and the gated potassium ion channels open. Positively charged potassium ions flood out. This makes the inside of the membrane once again −70 mV (with respect to the outside), and another action potential can be generated.

In practice, the potassium ions flood out just a little too fast and reduce the membrane potential to −80 mV. This 'undershoot' is called **hyperpolarisation**. It is quickly restored to the resting potential of −70 mV.

The events of depolarisation and repolarisation can be shown on a graph of membrane potential against time (see Figure 17). It is important to realise that this graph shows changes that occur *over a short period of time* at *one place* in the axon.

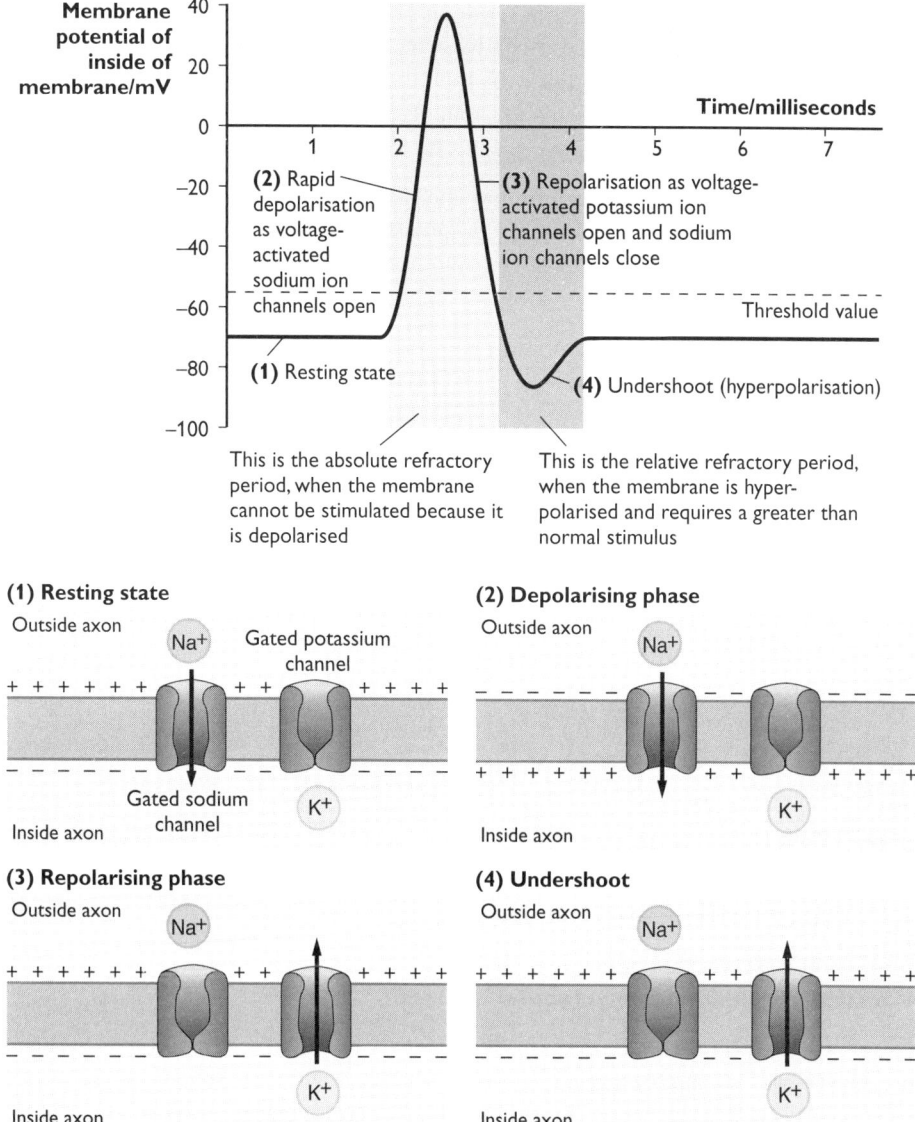

Figure 17 The depolarisation and repolarisation cycle

There is a period of time following the initiation of one action potential during which another action potential cannot be generated. This is called the **refractory period** and includes:
- the absolute refractory period (the periods of depolarisation and repolarisation)
- the relative refractory period (the period of hyperpolarisation)

The duration of the refractory period limits how many action potentials can be initiated per second. This, in turn, limits the number of nerve impulses that can be transmitted along a neurone per second.

Key facts you must know and understand

All action potentials are initiated in the same way; however, all neurones do not have the same structure. There are two basic kinds of neurones:
- myelinated neurones, which have a myelin sheath (Figure 14, page 23)
- non-myelinated neurones, which do not have a myelin sheath (Figure 18)

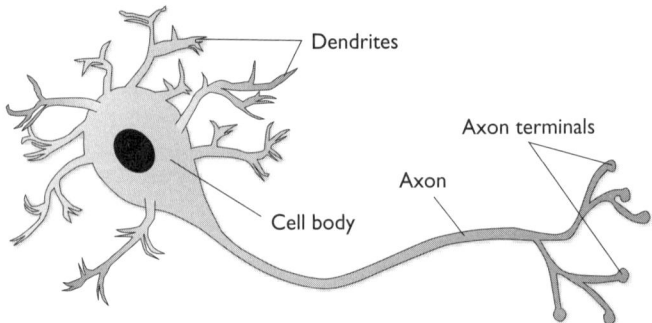

Figure 18 A non-myelinated neurone

The action potential is propagated differently along the two types of neurone.

Propagating action potentials along an axon

Key concepts you must understand

Propagation along non-myelinated neurones

Gated ion channels can exist in three states:
- open — appropriate ions can pass through
- closed but activated — nothing can pass, but the gate will open when stimulated
- closed and inactivated — nothing can pass and the gate will not open when stimulated (think of it as being 'locked')

In the 'resting' state, all the gated ion channels are closed, but activated.

When an action potential is generated at one end of an axon, the depolarisation affects the axon membrane just ahead of it. This region now begins to depolarise and generate an action potential in this new region. The region where the action potential originated begins to repolarise.

The action potential in the new region now affects the axon membrane just ahead of it and this begins to depolarise...and so on down the axon.

The gated sodium ion channels in the repolarising region are inactivated as well as being shut. They only become activated again once the membrane is fully repolarised. By this time the action potential has moved a considerable way down the axon and cannot affect the area where the action potential originated. This explains why action potentials cannot move backwards to where the action potential originated.

Propagation of an action potential along a non-myelinated neurone is shown in Figure 19.

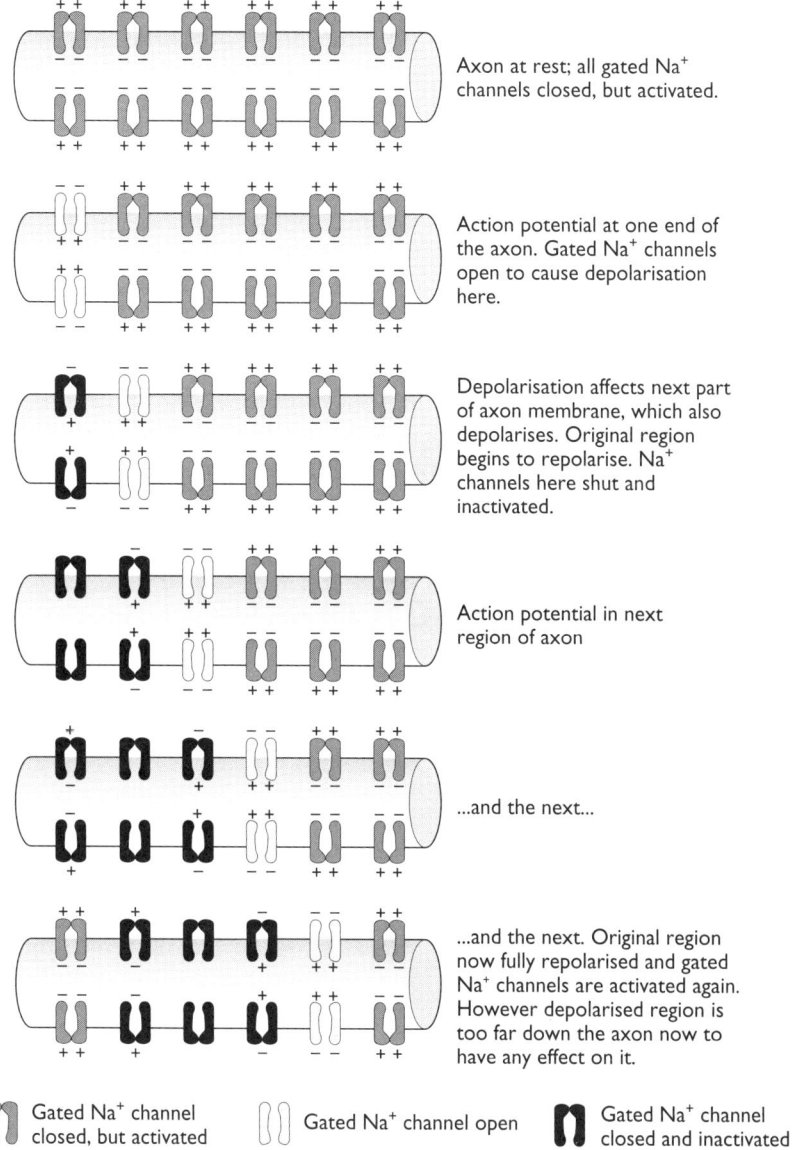

Figure 19 *Propagation of an action potential along a non-myelinated neurone*

Propagation along myelinated neurones

Myelinated neurones propagate action potentials by a method called **saltatory conduction** (this name comes from the Latin verb *saltare* — 'to jump').

In a non-myelinated neurone, each portion of the axon depolarises in turn and the action potential is propagated along the entire axon. In myelinated neurones, the action potential is only generated at the nodes of Ranvier —'gaps' in the myelin sheath. The nerve impulse 'jumps' from node to node and so is conducted *much* faster than in non-myelinated neurones (Figure 20).

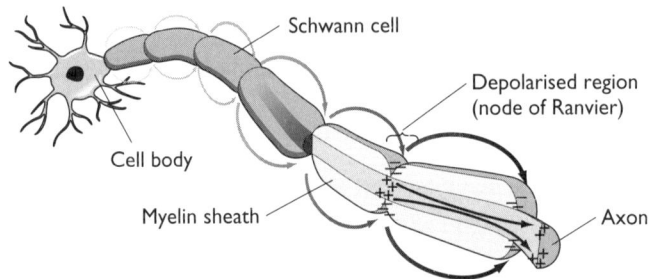

Figure 20 *Saltatory conduction in a myelinated neurone*

Transferring information about stimuli of different intensities

Whichever type of conduction of action potential takes place (saltatory or non-saltatory), all the action potentials generated have a value of +40 mV, regardless of the intensity of stimulation of the sense cells.

However, this is not true of generator potentials — a more intense stimulus results in a *larger* generator potential in the sense cell (a greater *amplitude* of the generator potential).

A larger generator potential does not result in bigger action potentials in the neurone, but in *more* action potentials per second (a greater *frequency* of action potentials).

So, a bright light will result in more action potentials along the optic nerve per second than a dim light. Similarly, a loud noise will result in more action potentials along the auditory nerve per second than a quiet sound.

Key facts you must know and understand

The speed of transmission of an action potential is affected by several factors. These include:
- temperature — like all physiological processes, an increase in temperature increases the rate of the process, up to the point where enzymes and transport proteins begin to denature
- diameter of the neurone — neurones with a large diameter conduct impulses faster than those with a narrow diameter
- myelination — myelinated neurones conduct impulses faster than non-myelinated neurones

Structure of the synapse

Key facts you must know and understand

A synapse is the structure that allows information to be passed from:
- one neurone to another
- a sense cell to a neurone
- a neurone to a muscle cell (this is often called a neuromuscular junction)

It is not simply the gap between two cells; this is the **synaptic cleft**.

Where information is transmitted between two neurones:
- the neurone that transmits information across the synaptic cleft is the **pre-synaptic neurone**
- the neurone that receives the information is the **post-synaptic neurone**

Transmission across the synaptic cleft occurs when neurotransmitters are secreted from vesicles in the pre-synaptic neurone into the synaptic cleft.

Synapses may be named according to the specific neurotransmitter that is secreted:
- At a **cholinergic synapse**, **acetylcholine** is the neurotransmitter.
- At an **adrenergic synapse**, **adrenaline** is the neurotransmitter.

The structure of a cholinergic synapse is shown in Figure 21.

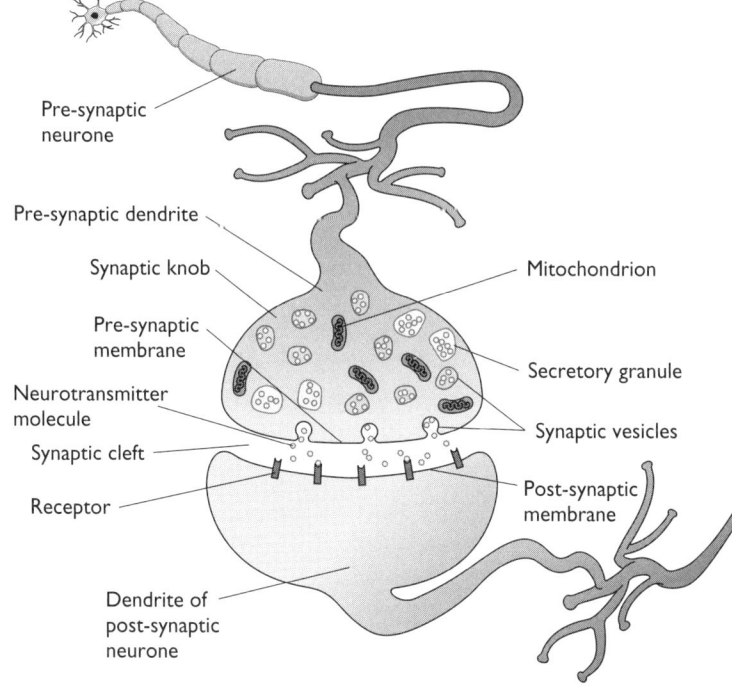

Figure 21 Structure of a cholinergic synapse

Transmission across the synapse

Key concepts you must understand

Synapses can be **excitatory** or **inhibitory**.

Different neurotransmitters are secreted at excitatory synapses and inhibitory synapses:
- At an excitatory synapse, once the neurotransmitter (e.g. acetylcholine) has been secreted, it binds to receptors on the post-synaptic membrane and *decreases* the membrane potential (makes it *less* negative). This makes it likely that the threshold will be reached and that an action potential will be initiated in the post-synaptic neurone.
- At an inhibitory synapse, once the neurotransmitter (e.g. GABA — gamma amino butyric acid) has been secreted, it binds to receptors on the post-synaptic membrane and *increases* the membrane potential (makes it *more* negative). This makes it *less* likely that the threshold will be reached and *less* likely that an action potential will be initiated in the post-synaptic neurone.

Usually, several neurones (not just two) synapse together. Some of these synapses may be excitatory and others inhibitory. The effect on the post-synaptic neurone depends on the combined effect of the neurones that release neurotransmitter at the same time. This is another example of **summation**. It is called **spatial summation** (Figure 22) because it is due to the combined effects of neurotransmitter at different places on the post-synaptic membrane, at the same time.

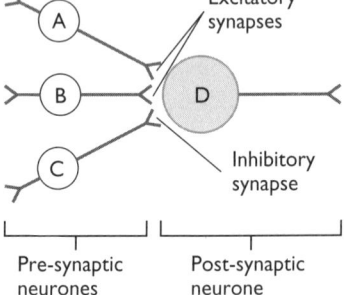

Figure 22 Spatial summation

The effect on the post-synaptic membrane depends on the strength of stimulation from each pre-synaptic neurone and the combination of stimulations. Some possible outcomes are shown in the table below (the letters refer to Figure 22).

Action potential in pre-synaptic neurones	Action potential in neurone D	Reasons
A	✗	Sub-threshold stimulation — voltage-activated sodium channels are not opened
B	✗	Sub-threshold stimulation — voltage-activated sodium channels are not opened
A + B	✓	The summation of effects of neurotransmitter of both excitatory synapses exceeds the threshold
A + C	✗	The summation of effects of excitatory and inhibitory synapses results in little change in the membrane potential — the threshold is not reached

Another type of summation is **temporal summation**. In temporal summation, several impulses arrive at the synapse from the same neurone in quick succession. Each depolarises the post-synaptic membrane a little more until threshold is reached.

Transmission at synapses only occurs in one direction (from pre-synaptic neurone to post-synaptic neurone), because:
- only the pre-synaptic neurone has synaptic vesicles containing neurotransmitter molecules
- only the post-synaptic membrane has receptor proteins for the neurotransmitter molecules
- there is a concentration gradient of neurotransmitter from pre-synaptic membrane to post-synaptic membrane

Key facts you must know and understand

The events that occur at excitatory and inhibitory synapses are summarised in the flowchart in Figure 23.

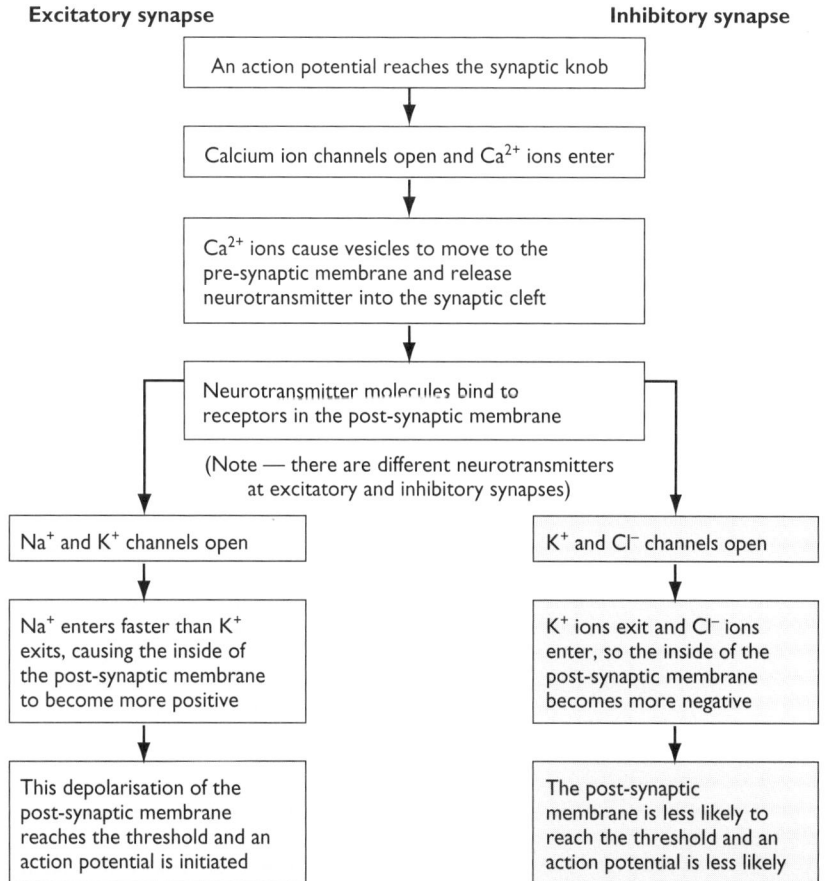

Figure 23 Differences between an excitatory and an inhibitory synapse

The organisation of neurones in reflex arcs

Key concepts you must understand

Reflex arcs are structures that allow reflex actions to be carried out. In a reflex action, the same stimulus always produces the same response.

A reflex arc must, therefore, always pass impulses along the same neural pathway. It always:
- receives information from a specific sense cell
- transfers information to the same specific effector

There are, broadly speaking, two main kinds of reflex actions:
- **Somatic reflexes** involve our special senses (eyes, ears, pressure detectors etc.) and produce a response by a muscle — for example, the 'withdrawal-from-heat' reflex.
- **Autonomic reflexes** involve sensors in internal organs and produce responses in internal organs — for example the control of heart rate and breathing rate.

Key facts you must know and understand

Many reflex arcs are built from three neurones:
- a sensory neurone
- a relay/inter-neurone
- a motor neurone

Figure 24 shows how they are organised in a typical somatic reflex arc.

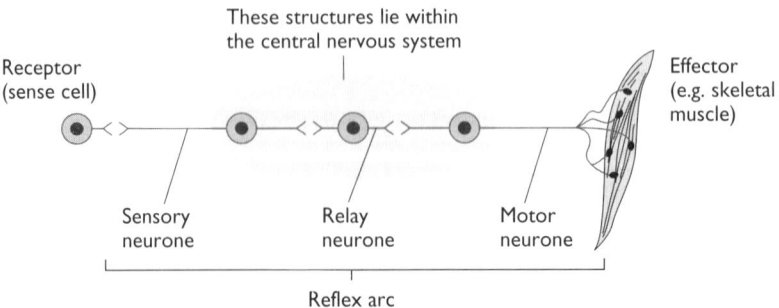

Figure 24 The structures in a somatic reflex arc

Reflex arcs in which the synapses with the relay neurone occur in the brain are those that control **cranial reflexes**; those in which the synapses occur in the spinal cord control **spinal reflexes**. Withdrawing your hand from a hot object is an example of a spinal reflex.

As was noted in the section on behaviour, many reflex actions are protective. A good example is the withdrawal reflex (see Figure 25).

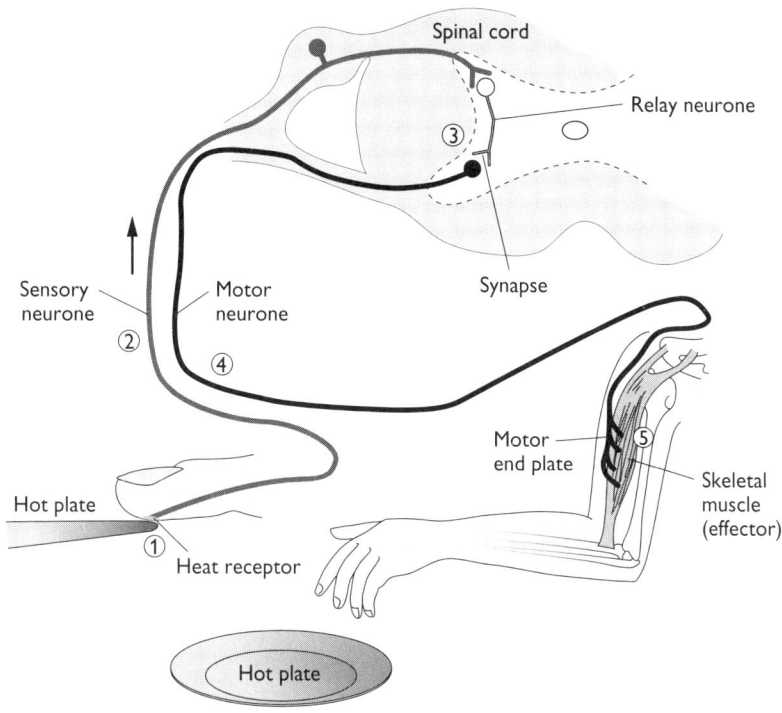

Figure 25 The withdrawal reflex

The sequence of events in the withdrawal-from-heat reflex is as follows:

(1) The heat receptor is stimulated by the hot plate and a generator potential results.

(2) The generator potential initiates an action potential in the sensory neurone, which is transmitted along the neurone.

(3) At the synapse with the relay neurone, neurotransmitter is released and initiates an action potential in the relay neurone.

(4) This is repeated at the synapse between the relay neurone and the motor neurone and an action potential is transmitted along the motor neurone.

(5) At the motor end plate, the action potential stimulates the contraction of the skeletal muscle, causing the automatic withdrawal of the hand from the hot plate.

Autonomic reflexes often occur in pairs. One is controlled by the sympathetic branch of the ANS, the other by the parasympathetic branch. The control of heart rate (Figure 26) is an example of autonomic reflex control involving two reflex arcs:
- The reflex arc involving the sympathetic branch of the ANS increases heart rate.
- The reflex arc involving the parasympathetic branch of the ANS decreases heart rate.

Notice the antagonistic action of the sympathetic and parasympathetic branches.

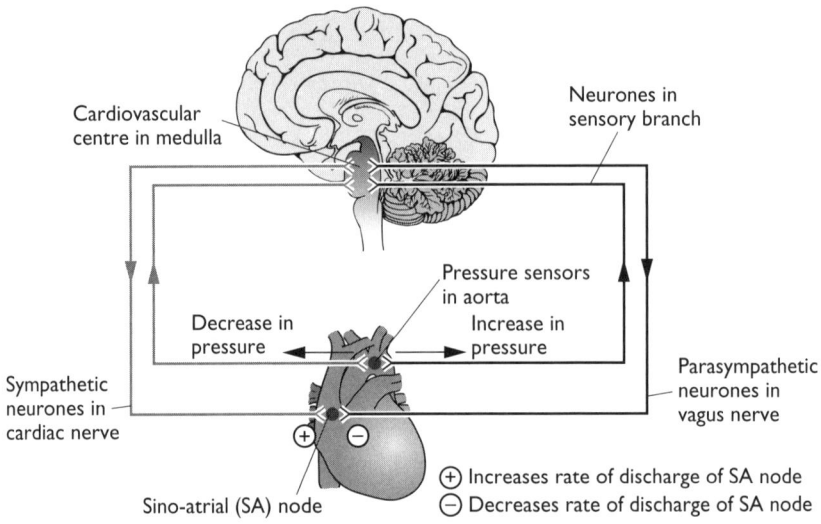

Figure 26 How reflex arcs in the ANS control heart rate

Sympathetic control	Parasympathetic control
Initiated by a decrease in blood pressure in the aorta	Initiated by an increase in blood pressure in the aorta
Impulses pass to the cardiovascular centre and then to the sino-atrial (SA) node in the heart via neurones in the cardiac nerve	Impulses pass to the cardiovascular centre and then to the SA node in the heart via neurones in the vagus nerve
Noradrenaline is released from endings of sympathetic neurones in the SA node, which increases activity of the node and so increases heart rate	Acetylcholine is released from endings of parasympathetic neurones in the SA node, which reduces activity of the node and so reduces heart rate
Other branches of the cardiac nerve affect the ventricle muscle and increase stroke volume	Other branches of the vagus nerve affect the ventricle muscle and reduce stroke volume

Control by hormones

Key concepts you must understand

Hormones are chemicals produced by **endocrine glands** that target particular cells in other parts of the body. They circulate round the body dissolved in the blood plasma. Their **target cells** (in **target organs**) have specific **receptor proteins** on or in them. The tertiary structure of these proteins has a binding site that has a shape *complementary* to only one hormone (see Figure 27).

> **Tip** Remember that the shapes are complementary, like an egg cup and an egg. Two objects with the same shape (such as two eggs) cannot fit together.

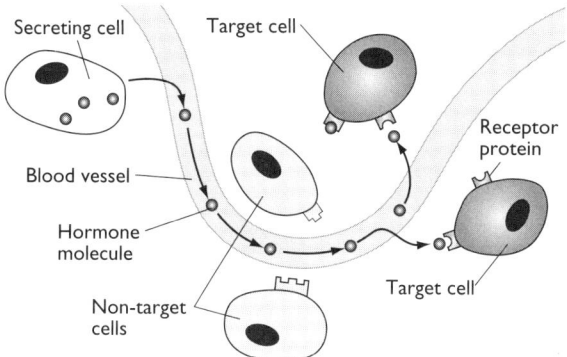

Figure 27 How hormones target specific cells

Hormones activate processes in cells in one of two main ways:
- Steroid hormones (for example the male and female sex hormones) are small, lipid-soluble molecules that can pass freely through plasma membranes and bind with receptors in the cytoplasm. The receptor–hormone complex then moves into the nucleus, binding with and activating specific genes.
- Non-steroid hormones (for example insulin and glucagon) bind to receptors on the surface of the cell. This activates a molecule in the membrane called a 'G-protein', which activates specific enzymes that produce specific metabolic effects.

Key facts you must know and understand

Hormonal control is different from nervous control in several ways. When compared with nervous control, hormonal control of responses is:
- slower, because the hormone is secreted into, and travels in, the bloodstream to reach its target cells
- longer lasting, because the effect persists for as long as the hormone remains bound to its receptor in or on the target cell
- more general in its effects, because hormones often have target cells in different regions of the body — some hormones affect nearly all cells

Prostaglandins, histamine and the inflammatory response to injury

Key concepts you must understand

Prostaglandins and **histamine** are 'chemical messengers' like hormones, but, unlike hormones, they act locally on the cells that produce them and other cells in the immediate area.

The inflammatory response to injury has evolved to:
- increase the blood flow to an infected or damaged area
- increase the migration of phagocytic white blood cells to the area

- allow phagocytic white blood cells to leave capillaries more easily
- promote the clotting of blood and repair to the damaged area

Key facts you must know and understand

The four characteristics of an inflammatory response are:
- redness
- warmth
- pain
- swelling

Prostaglandins are released from damaged cells at the site of injury and cause arterioles to dilate. This allows more blood to flow to the area (causing the redness and warmth) and more phagocytic white blood cells to migrate to the area. The

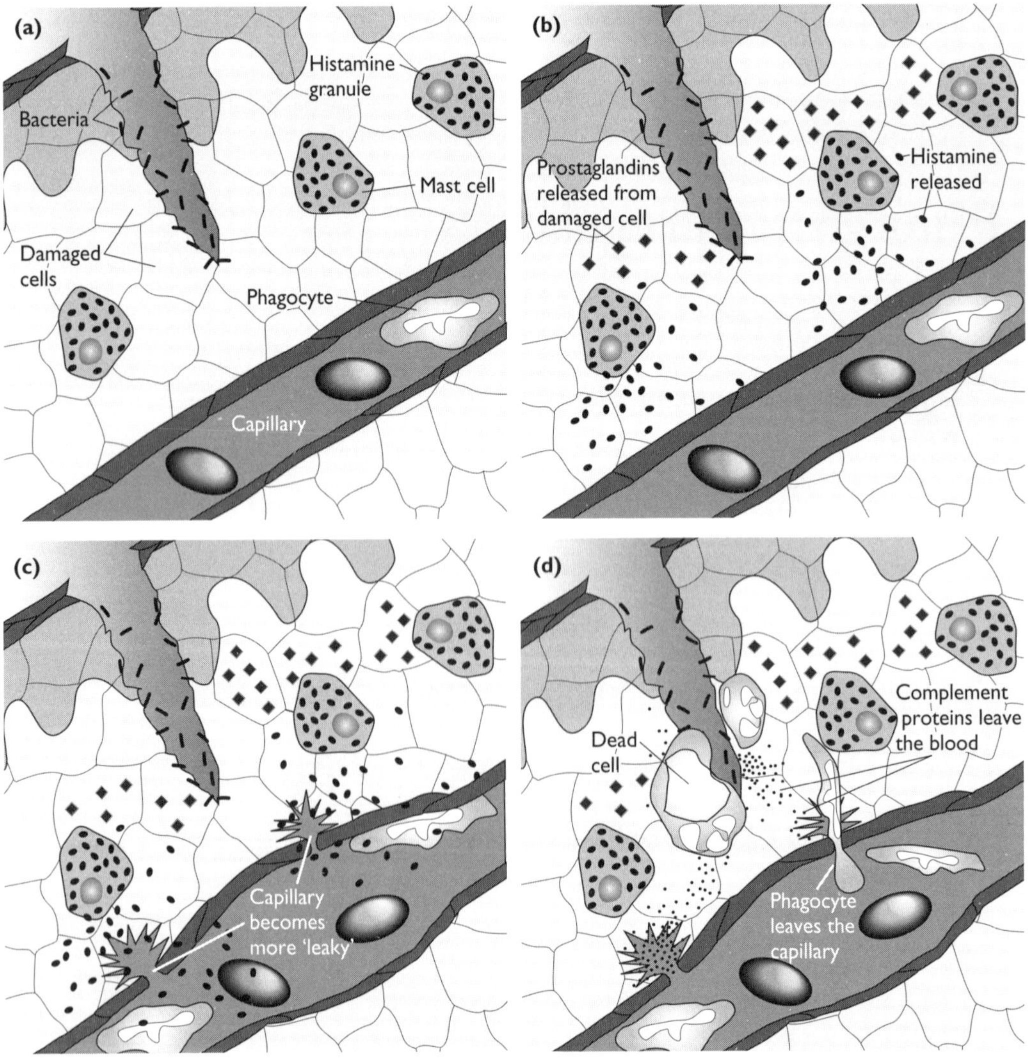

Figure 28 How prostaglandins and histamine bring about the inflammatory response

increased blood flow is, indirectly, the cause of the pain, because the increased volume of blood increases pressure on nerve endings and pressure receptors.

Prostaglandins also promote clotting of the blood. This minimises both blood loss and the entry of microorganisms.

Histamine is produced by **mast cells** in the area. Histamine acts on the capillary walls, allowing these to dilate to accommodate the increased blood flow and to become more 'leaky' than usual, which allows:
- more plasma to leave the blood
- complement proteins (important in the immune response) to leave the blood
- phagocytes to leave the blood more easily

The role of prostaglandins and histamine in the inflammatory response is shown in Figure 28.

The liquid in the tissues together with dead phagocytes containing the bacteria and/ or dead cells they have engulfed, eventually escape from the inflamed site as 'pus'.

Plant growth regulators

Key concepts you must understand

Auxins are the best known plant growth regulators and were the first to be discovered.

The discovery of the action of auxins: how scientists take forward the research of other scientists

Much of the research into the action of auxins has been carried out using coleoptiles.

> **Tip** A coleoptile is *not* a stem, so don't call it one. It is the curved sheath that covers the emerging shoot of a seedling of a cereal.

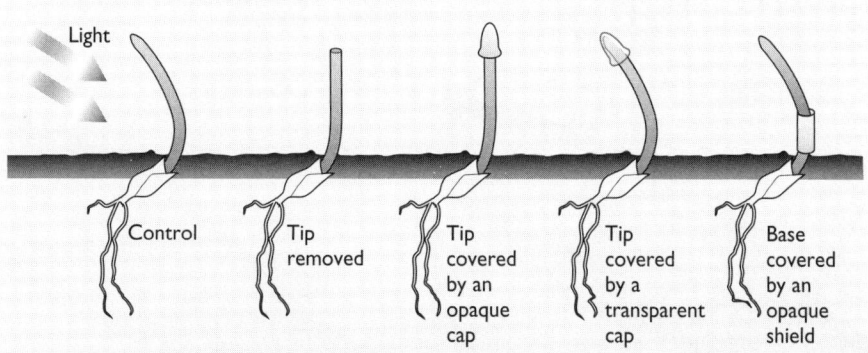

Figure 29 The research by Charles and Francis Darwin

The Darwins concluded that
- detection of the stimulus occurs in the tip of the coleoptile
- a region some distance behind the tip brings about the response
- there was communication between the two, probably chemical in nature

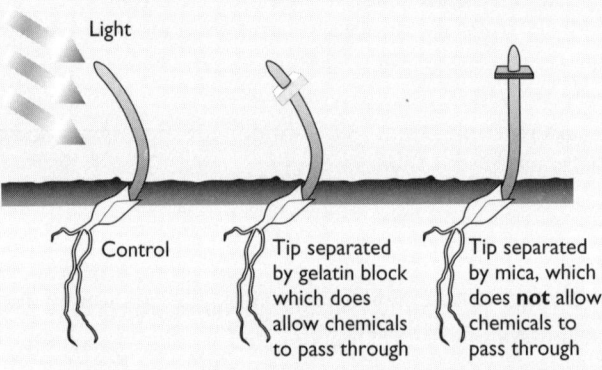

Figure 30 The research by Boysen-Jensen

Boysen-Jensen concluded that the communication between detection and response was chemical in nature.

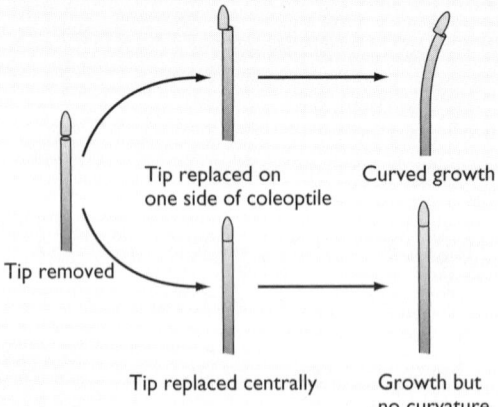

Figure 31 The research by Paal

Paal concluded that his results provided evidence that curved growth was due to the 'chemical messenger' accumulating on one side of the stem.

Conclusions to the experiments as shown have to be tentative because:
- there is no control to show the effect of not replacing the cut tip
- there is no measure of a chemical accumulating

Went investigated the effects of different concentrations of the 'chemical messenger' on the growth of coleoptile tips.

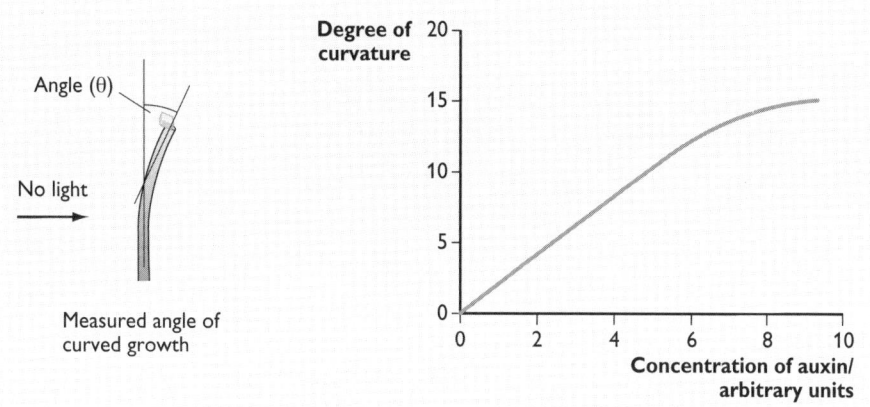

Figure 32 The research by Went

Went concluded that increasing concentrations of the 'chemical messenger' caused increased growth, up to a maximum.

Since then, the auxin has been isolated and purified and shown to have exactly the same effects as the 'chemical messenger' first predicted by the Darwins.

Biologists believe that auxins act on growth genes, turning them on and stimulating both cell division and cell elongation (Figure 33).

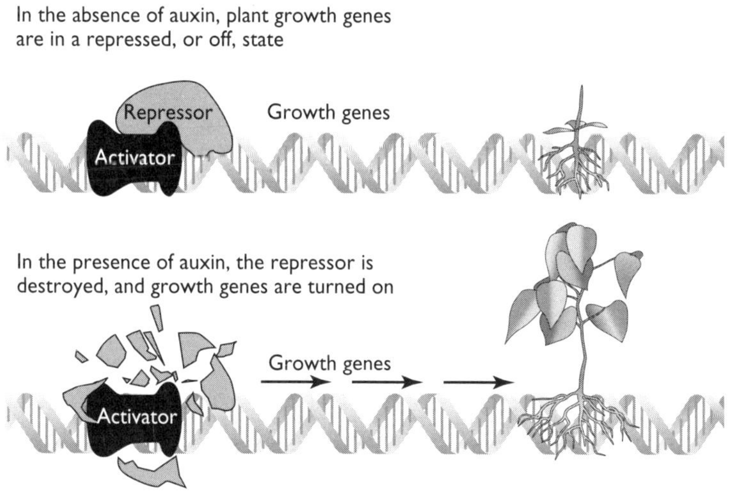

Figure 33 Auxins act by inactivating repressor substances that block the activation of growth genes in plants

The growth towards light from one side is a result of the auxin being redistributed to the shaded side of the shoot (Figure 34).

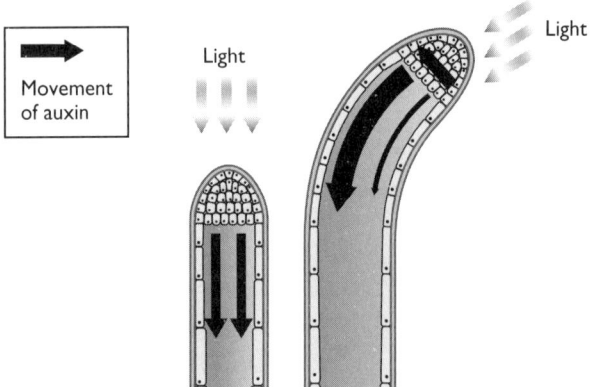

Figure 34 The phototropic response of a shoot. Auxins move away from the lit side of a shoot and accumulate in the shaded side. This side grows faster as a result, causing a curvature towards the light.

The **gravitropic response** is also controlled by auxins. However, auxins inhibit growth in roots. Growth is brought about by either the absence, or low concentrations of auxins.

If a seedling is placed horizontally and left for a period, the root will grow downwards and the shoot upwards.

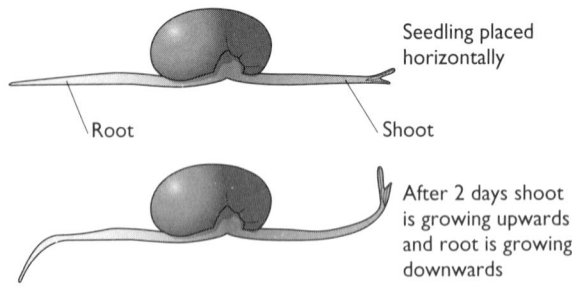

Figure 35 Investigating how a root and shoot respond to gravity

This happens because, in both root and shoot, auxin is redistributed to the lower side.
- In the shoot, auxin stimulates growth. Therefore, cells on the lower side grow faster than those on the upper side. This causes an upward curvature.
- In the root, auxin inhibits growth. Therefore, cells on the lower side grow more slowly than those on the upper side. This causes a downward curvature.

What the examiners could ask you to do

- Explain any of the key concepts.
- Recall and show understanding of any of the key facts.

- Interpret information about responses and deduce whether they are controlled by hormones or by nerves.
- Compare and contrast nervous and hormonal control.
- Interpret diagrams showing changes in membrane potential.
- Interpret diagrams of synapses.
- Explain how one-way transmission is ensured along axons and across synapses.
- Interpret diagrams of nerve networks showing summation at synapses.
- Explain the significance of the specificity of receptor proteins in the post-synaptic membrane and in the target cells of hormones.
- Relate stimulus intensity to the size of generator potentials and to the frequency of action potentials.
- Use your knowledge of the tertiary structure of proteins, the structure of plasma membranes, transport across membranes, aerobic respiration and the circulatory system to explain some of the events that take place in nervous and in hormonal control.
- Comment on the reliability and/or validity of investigations into nervous and chemically mediated responses, bearing in mind:
  - the numbers of organisms used and/or repeats carried out
  - possible flaws in the technique or in the apparatus
  - overlap of error bars in graphs of results
  - the extent to which appropriate controls were carried out

Skeletal muscle: a key effector in mammals

The structure of skeletal muscle

Key concepts you must understand

Muscle tissue contains two fibrous proteins (**actin** and **myosin**) that interact to bring about contraction.

Muscles (such as the biceps, triceps and rectus muscles) are organs. They contain several tissues, each important to the functioning of the muscle. These include:
- muscle tissue, to provide the force for contraction
- blood vessels, to carry the blood (arteries and veins are also organs)
- blood, bringing oxygen and glucose for respiration
- sense cells, to detect the degree of 'stretch' in the muscle
- sensory neurones, to carry impulses from the stretch receptors
- motor neurones, to initiate contraction of the muscle
- connective tissue, to 'pack' the groups of muscle fibres

A2 Biology

> **Tip** Be sure you are able to distinguish between muscle tissue and muscles. Muscle tissue is just one of the tissues (they are all listed above) that are found in a muscle (such as biceps or triceps).

Key facts you must know and understand

A muscle is made from bundles of fibres. Each fibre contains many **myofibrils** made from thick and thin **myofilaments**. A single 'contracting unit' of a myofibril is called a **sarcomere**. The structure of a muscle is shown in Figure 36.

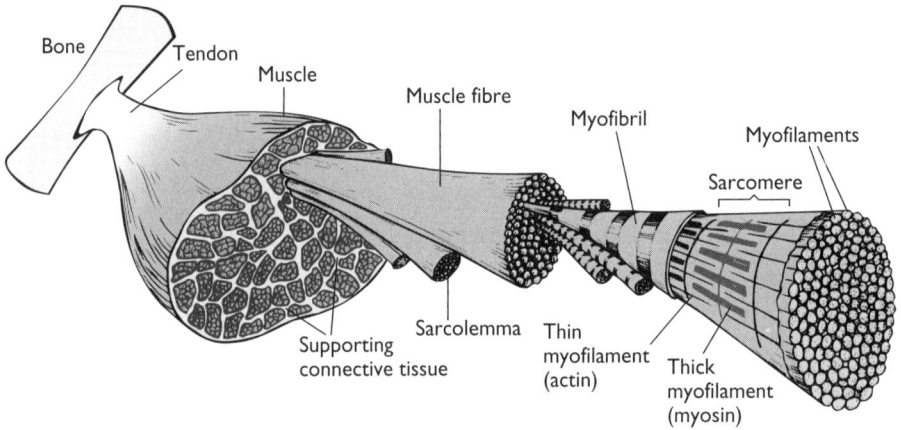

Figure 36 The structure of skeletal muscle

How skeletal muscle moves bones: the sliding filament theory

Key concepts you must understand

Skeletal muscle can only exert a force when it *contracts*; it can therefore only *pull* a bone. To reverse the movement, a second muscle is needed. Skeletal muscles are, therefore, always found in **antagonistic pairs** — for example the biceps and triceps muscles that flex (bend) and extend (straighten) the arm.

Skeletal muscle contracts as a result of the interaction between two proteins, actin and myosin, found throughout the muscle.

The generally accepted theory of muscle contraction is called the **sliding filament theory** (see Figure 37). According to this theory, filaments of actin and myosin slide over one another, shortening each sarcomere. As a result, each myofilament becomes shorter and so the whole muscle becomes shorter.

AQA Unit 5

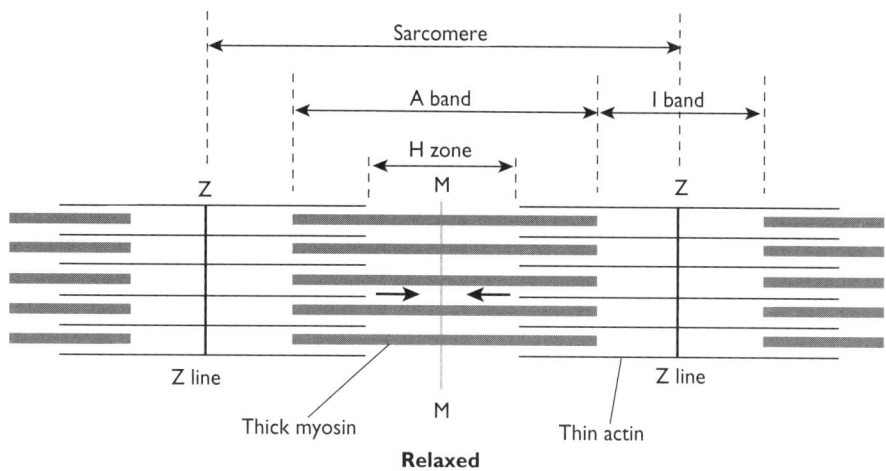

Contracted
(note that there is now no I band or H zone)

Figure 37 Shortening of a sarcomere according to the sliding filament theory

Key facts you must know and understand

An outline of the mechanism
- The ends of the myosin filaments are bulbous and are called **myosin heads**.
- They bind with **binding sites** on the actin filaments.
- As they bind, the heads tilt, and move the actin filaments.
- The myosin heads are then released and return to their original positions.
- They bind to other sites further along the actin filament and repeat the tilting and moving.

> **Tip** There is quite a lot of detail to explain how this takes place. Make sure that you understand the outline fully.

The stages in muscle contraction
When the muscle is relaxed:
- molecules of **tropomyosin** block the binding sites, preventing the myosin heads from binding
- each myosin head is in a 'resting position' and is bound to a molecule of ATP

Neurones synapse with myofibrils at **neuromuscular junctions** (Figure 38). Nerve impulses initiate contraction when they release neurotransmitters into the junction.

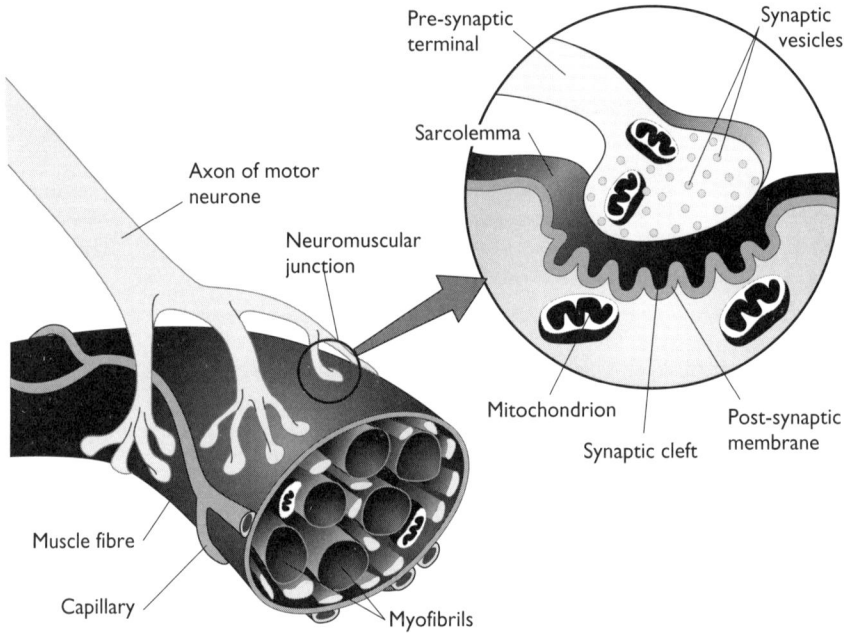

Figure 38 A neuromuscular junction

Muscle contraction is caused in the following way:
- Nerve impulses arrive at the neuromuscular junction.
- Neurotransmitter molecules are released and cross the synaptic cleft.
- This causes calcium ions to be released around the actin molecules.
- These ions cause the tropomyosin molecules to move and expose the binding sites on the actin molecules.
- At the same time, the ATP molecules are hydrolysed, to form ADP and inorganic phosphate (which is released). The energy released is transferred to the myosin heads.
- The myosin heads bind to the exposed binding sites on the actin molecules (sometimes referred to as **cross-bridge formation**), releasing the ADP.
- The energy previously transferred to the myosin heads now moves the heads and with them the actin filaments.
- Another ATP molecule attaches to each myosin head, which now detaches from the actin and binds to a binding site further along the actin molecule.
- Provided nervous stimulation continues to cause calcium ions to be released, the cycle continues. Otherwise, the muscle returns to the relaxed state.

This sequence of events is shown diagrammatically in Figure 39.

AQA Unit 5

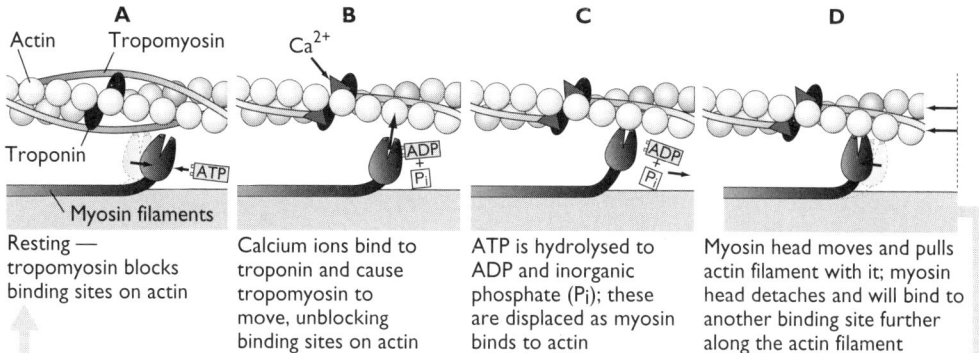

Figure 39 How the sliding filament theory explains muscle contraction

How muscles transfer the energy they need to contract

The ATP needed for muscle contraction is supplied in three ways:
- aerobic respiration
- anaerobic respiration
- the **phosphocreatine–ATP system**

Molecules of phosphocreatine (CP) store energy in a similar way to ATP. Energy is used to attach a phosphate group to a creatine molecule. When phosphocreatine is hydrolysed, the energy is released again.

During exercise, energy is released in the following ways:
- At the start of exercise, the small store of ATP held in muscles is hydrolysed to ADP and inorganic phosphate; this is used up quickly.
- CP is then hydrolysed to creatine and phosphate, which releases the energy needed for the resynthesis of ATP. This releases energy quickly, but is also used up quickly. It is the main source of ATP in 'explosive' events.
- For longer periods of exercise, ATP is resynthesised by either aerobic or anaerobic respiration or by a combination of the two. As the duration of the exercise increases and/or the intensity decreases, more ATP is produced by aerobic respiration.

Types of muscle fibre

Key facts you must know and understand

There are two types of muscle fibre:
- **slow-twitch fibres**, which release energy principally by aerobic respiration
- **fast-twitch fibres**, which release energy principally by anaerobic respiration

Slow-twitch fibres are adapted to releasing energy aerobically over a sustained period. They have:
- many mitochondria in each fibre

- high concentrations of the enzymes that regulate the Krebs cycle; most reduced NAD and reduced FAD molecules (that release electrons into the electron transport chains of the inner mitochondrial membrane) are produced in the Krebs cycle
- a more extensive capillary network than fast fibres (allowing more oxygen to be delivered)
- a high concentration of **myoglobin**, which stores oxygen
- a relatively slow contraction rate with less force than fast-twitch fibres
- a high resistance to fatigue

Fast-twitch fibres are adapted to releasing energy anaerobically over a short period of time. Fast-twitch fibres:
- have fewer mitochondria than slow-twitch fibres
- have high concentrations of the enzymes that control glycolysis (and low concentrations of the Krebs cycle enzymes)
- have a higher concentration of ATPase than slow-twitch fibres so that a lot of ATP can be hydrolysed quickly
- have a higher contraction rate, with more force than slow-twitch fibres
- have low concentrations of myoglobin
- have a lower resistance to fatigue than slow-twitch fibres

What the examiners could ask you to do

- Explain any of the key concepts.
- Recall and show understanding of any of the key facts.
- Interpret diagrams of relaxed and contracted skeletal muscle fibres.
- Interpret diagrams showing the different types of muscle fibres in a muscle and relate these to the abilities of the person from whom the muscle came.
- Use your knowledge of aerobic and anaerobic respiration, the structure and properties of ATP, transmission across synapses and the oxygen dissociation curve of haemoglobin to explain some of the events in muscle contraction.
- Comment on the reliability and/or validity of investigations into muscle contraction (possibly linked to exercise), bearing in mind:
  - the numbers of organisms or people used and the number of repeats carried out
  - possible flaws in the technique or in the apparatus
  - overlap of error bars in graphs of results
  - the extent to which appropriate controls were carried out

Homeostasis and feedback systems

Key concepts you must understand

Homeostasis describes a condition in which the internal environment is allowed to change within only very narrow limits.

The cells of the body are surrounded by **tissue fluid** which has a near constant:
- temperature
- pH
- water potential

Maintaining a constant internal environment allows cellular enzymes to work in their optimum conditions and, therefore, carry out metabolic processes with maximum efficiency.

These factors are maintained within their limits by **negative feedback systems** (Figure 40). Such systems detect when the value of a particular factor falls outside its 'preset range' and activates mechanisms to bring it back within that range.

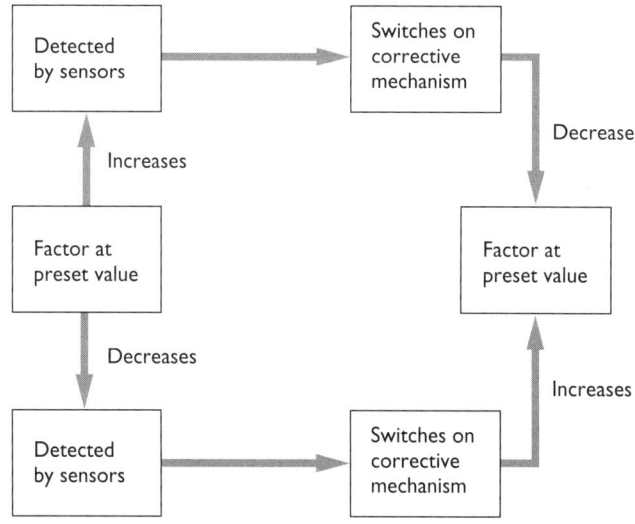

Figure 40 The principle of negative feedback

Some systems in the body operate using **positive feedback**. In such systems, a change is detected and the change is then amplified (made larger). The system only 'reverts to normal' when the stimulus is removed.

Key facts you must know and understand

Regulating core body temperature

Animals can be placed into two groups on the basis of how they regulate their core body temperatures:
- **Endotherms** use physiological processes and behavioural methods to regulate their core temperature.
- **Ectotherms** use only behavioural methods to regulate their core temperature.

> **Tip** Be aware that the classification into two types is not absolute. Some reptiles have limited ability to control their temperature by physiological means.

> **Tip** Do *not* refer to the two types as 'warm-blooded' and 'cold-blooded'. These terms are extremely inaccurate.

Core body temperature will remain constant if heat losses balance heat gains.

Humans gain heat in the following ways:
- radiation from other objects
- eating warm food
- respiration — much of the energy released in respiration is released as heat; very active organs such as the liver and active skeletal muscle generate large amounts of heat

Humans lose heat by:
- conduction — when we touch something cooler than our skin
- convection — when water vapour in our breath and sweat moves away from our body, carrying heat with it
- radiation — when heat is radiated from our body surface (occurs at all times)
- evaporation — when water in sweat turns to a vapour on our skin at least part of the heat to make this happen comes from the skin

Core body temperature is regulated by the **thermoregulatory centre** in the **hypothalamus**. This is divided into two regions:
- The **heat gain centre** activates measures that conserve heat and that increase heat generation.
- The **heat loss centre** activates measures that increase heat loss and decrease heat generation.

As the heat gain centre becomes active it inhibits the heat loss centre. The heat loss centre also inhibits the heat gain centre when it becomes active.

The hypothalamus receives information about body temperature from two sources:
- Sensors in the skin detect *changes* in the temperature of the skin.
- Sensors in the hypothalamus itself detect *changes* in the temperature of the blood flowing through it which reflects changes in the 'core' body temperature.

> **Tip** Notice that the sensors detect *changes* in the temperature. They are not thermometers – they do not detect the *actual* temperature.

Figure 41 summarises the role of the hypothalamus in controlling body temperature.

The hypothalamus 'orders' the regulatory processes through the **autonomic nervous system** (ANS). The two divisions once again act antagonistically.

The **sympathetic division** carries impulses from the heat gain centre to:
- increase metabolic rate and heat generation
- activate processes that reduce heat loss

The **parasympathetic division** carries impulses from the heat loss centre to:
- decrease metabolic rate and heat generation
- activate processes that increase heat loss

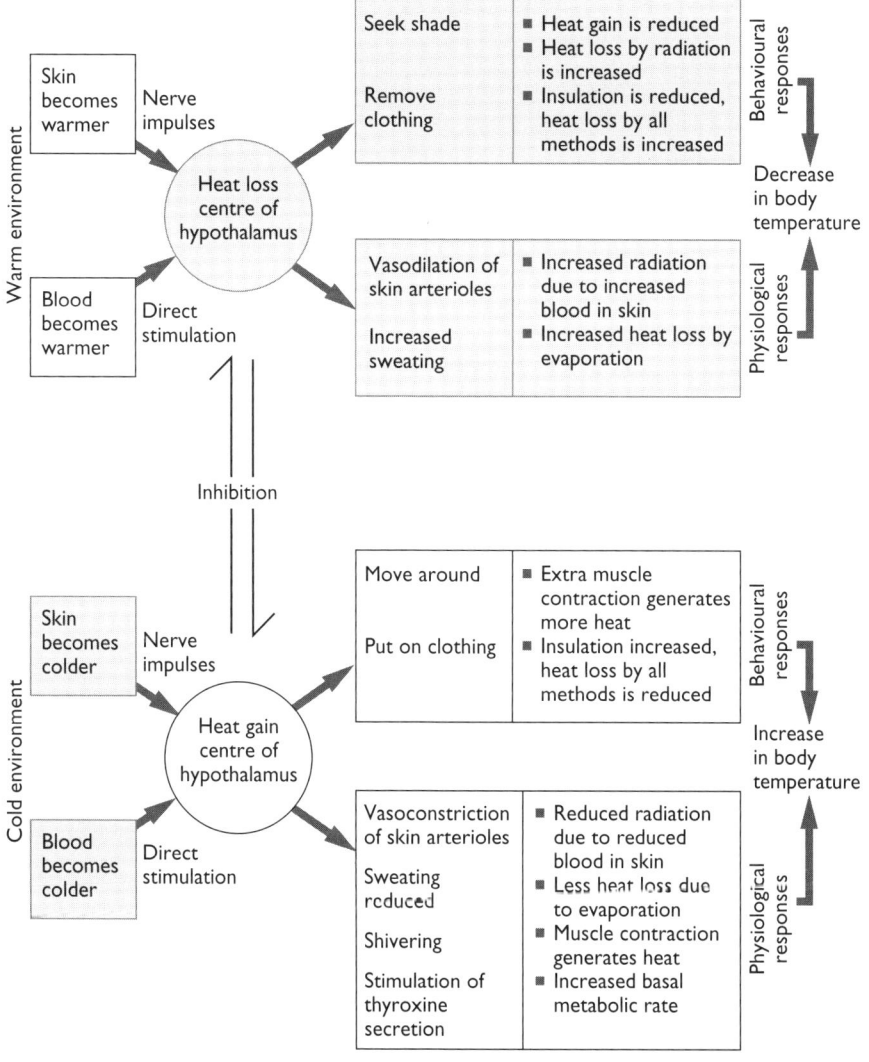

Basal metabolic rate is the rate of energy expenditure when a person is awake but resting, has not eaten for 12 hours and is comfortably warm.

Figure 41 Mechanisms involved in regulating core body temperature in humans

Controlling plasma glucose concentration

The islets of Langerhans in the pancreas contain two different types of secretory cell:
- α-cells, which secrete the hormone **glucagon**
- β-cells, which secrete the hormone **insulin**

Both these hormones activate enzymes (in the liver and skeletal muscle in particular) that influence carbohydrate metabolism.

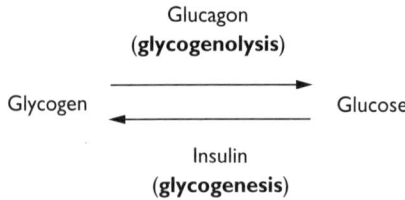

Tip The term 'glycogenesis' means 'making glycogen' (from glucose). 'Glycogenolysis' means 'splitting glycogen' (into glucose). Remember, 'lysis' always means 'splitting' – as in hydro*lysis*.

Figure 42 summarises the control of plasma glucose concentration by insulin and glucagon. Glucose concentration is returned to normal when it becomes too high (after a meal has been digested) or too low (some time after the absorption of glucose from a meal).

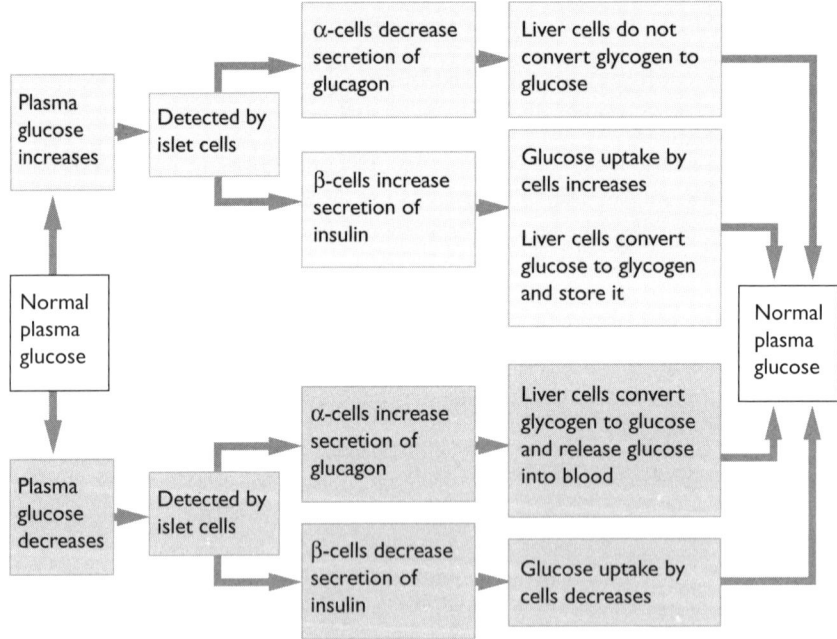

Figure 42 The role of insulin and glucagon in the regulation of plasma glucose concentration

Both hormones activate the interconversion of glucose and glycogen by:
- binding to a receptor protein in the plasma membrane of target cells in the liver
- activating a 'cascade system' through a 'second messenger molecule' (cyclic AMP; see Figure 43)

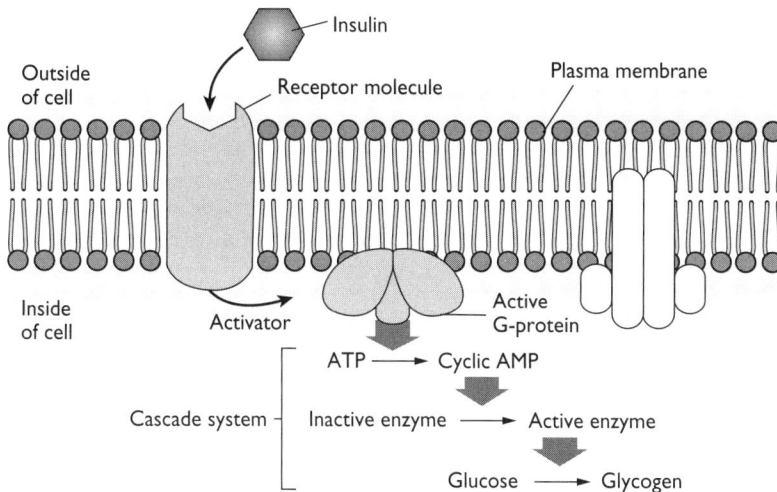

Figure 43 How insulin activates the conversion of glucose to glycogen in liver cells

Glucagon activates the conversion of glycogen to glucose in a similar way, but binds to a different receptor protein that activates a different enzyme.

Other hormones also influence plasma glucose concentration, but do not have the same regulatory function of insulin and glucose. These include:
- **adrenaline**, which increases the breakdown of glycogen to glucose and the release of glucose into the bloodstream to allow increased energy release in the muscles
- **thyroxine**, which increases the basal metabolic rate, so an increased rate of energy release is required
- **corticosteroids**, which promote reactions that result in the synthesis of glucose from non-carbohydrate sources: this is **gluconeogenesis** (see Figure 44).

> **Tip** To remember the term 'gluconeogenesis', break the word down. *Gluco* (= glucose), *neo* (= new) and *genesis* (= making) together mean 'making new glucose'.

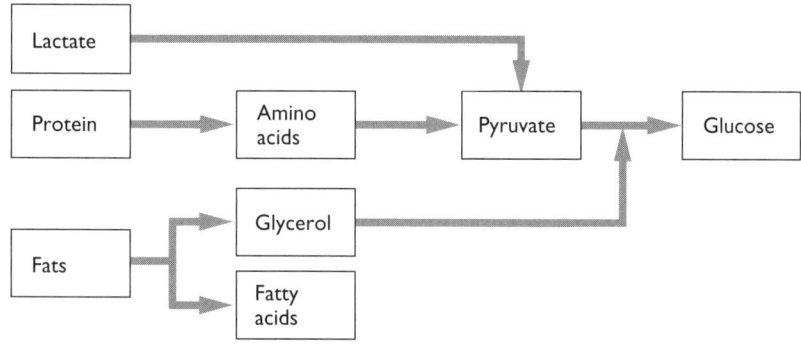

Figure 44 The main pathways of gluconeogenesis

Type 1 and type 2 diabetes

Diabetes is a condition in which insufficient insulin is secreted in response to high plasma glucose concentrations.
- **In type 1** diabetes, the immune system attacks the β-cells in the islets of Langerhans, so they cannot produce and secrete sufficient insulin.
- **In type 2** diabetes, the β-cells still produce some insulin (but less than normal) and the receptors in the plasma membranes of liver cells lose their sensitivity to the hormone.

Type 1 is sometimes called '**early-onset**' diabetes and type 2, '**late-onset**' diabetes.

Sustained **hyperglycaemia** (high plasma glucose concentrations) results in two key changes in the body:
- Not all the glucose filtered from the blood in the kidney is reabsorbed.
- Tissues become dehydrated as they lose water by osmosis to the blood (because of the decreased water potential of the plasma)

These changes are the cause of several of the symptoms of diabetes, as shown in Figure 45.

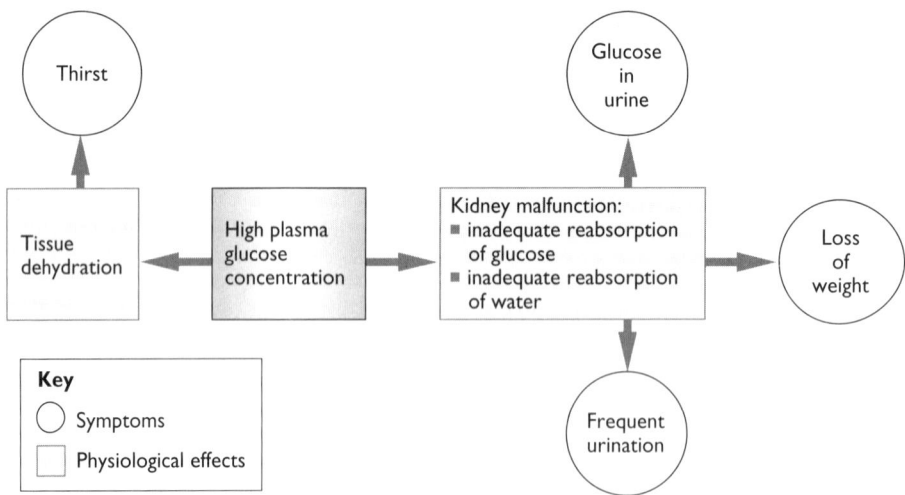

Figure 45 How hyperglycaemia causes the symptoms of diabetes

Type 1 diabetes is treated by regular injections of insulin.

Type 2 diabetes is treated by:
- a diet planned to avoid 'peaks' and 'troughs' in the concentration of plasma glucose
- an exercise programme to ensure that glucose in the plasma is used
- oral anti-diabetes medication to
  – reduce gluconeogenesis in the liver
  – enhance the production of insulin in the islets of Langerhans

Positive feedback systems

Positive feedback occurs during childbirth (see Figure 46), when contractions of the uterus stimulate the pituitary gland to secrete the hormone **oxytocin**. This stimulates stronger contractions, which stimulate increased secretion of oxytocin, which stimulates even stronger contractions of the uterus...and so on.

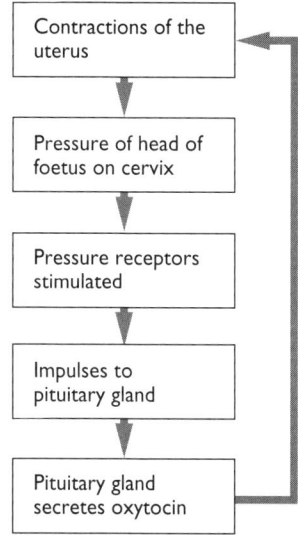

Stimulation of pituitary gland causes more oxytocin to be secreted, which increases the strength of the contractions

Figure 46 A flow chart showing positive feedback during childbirth

After birth, the stimulus of contractions of the uterus is removed, so oxytocin is no longer secreted and the 'system reverts to normal'.

Positive feedback sometimes occurs when the normal negative feedback systems fail to function properly. A good example of this is the progressive development of **hypothermia**.

Usually, when body temperature falls, the body responds by increasing its metabolic rate. This results in more energy being released. However, if this is not enough to offset the heat losses, core body temperature continues to fall. Beyond a certain point, further reduction in body temperature *reduces* the metabolic rate, so *less* heat is produced. This means that the core body temperature now reduces more quickly — so the metabolic rate reduces more quickly, and so on.

Positive feedback and negative feedback systems control mammalian oestrous cycles

The human menstrual cycle is one example of an oestrous cycle. During the cycle of approximately 28 days, an **oocyte** (immature egg cell) is released from one of the ovaries and the lining of the uterus is made ready for implantation, if the oocyte is fertilised. If fertilisation does not take place, the uterine lining is lost in menstruation.

The events in this cycle are under the control of four hormones (see Figure 47) — two gonadotrophic hormones and two steroid sex hormones.

Gonadotrophic hormones are secreted by the pituitary gland and target cells in the follicles in the ovaries:
- **Follicle stimulating hormone** (**FSH**) stimulates the development of a follicle in an ovary and causes it to secrete oestrogen.
- **Lutenising hormone** (**LH**) stimulates ovulation (the release of an oocyte) as well as stimulating the follicle cells to secrete progesterone.

Steroid sex hormones are secreted by cells in the ovary. They target the uterine lining as well as stimulating and/or inhibiting the secretion of gonadotrophic hormones:
- **Oestrogen** stimulates the development of the uterine lining, inhibits the secretion of FSH by the pituitary gland and stimulates the secretion of LH.
- **Progesterone** maintains the uterine lining and inhibits the secretion of both FSH and LH by the pituitary gland.

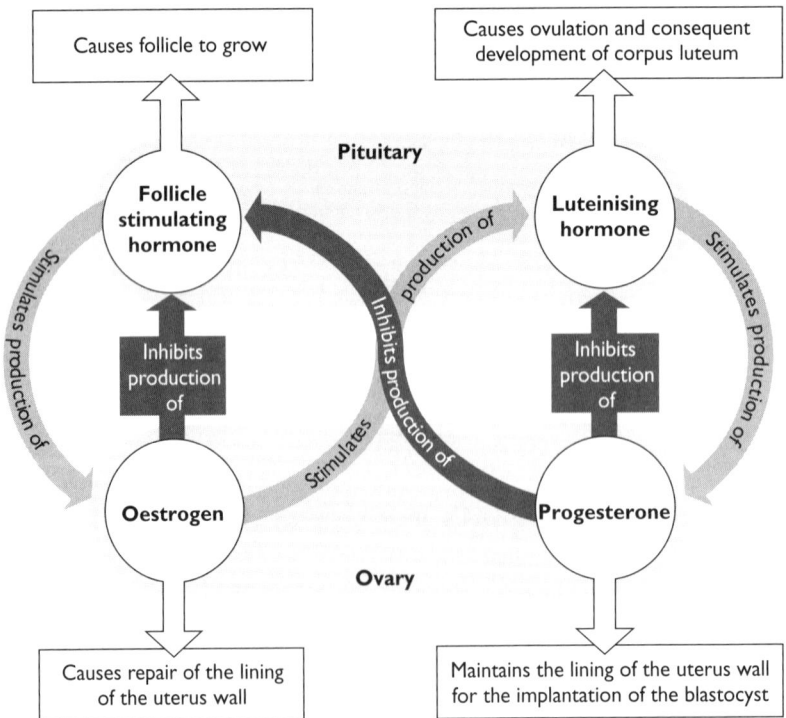

Figure 47 The interactions between the hormones controlling the human menstrual cycle

The changes in concentrations over the 28-day cycle can be represented graphically (Figure 48).

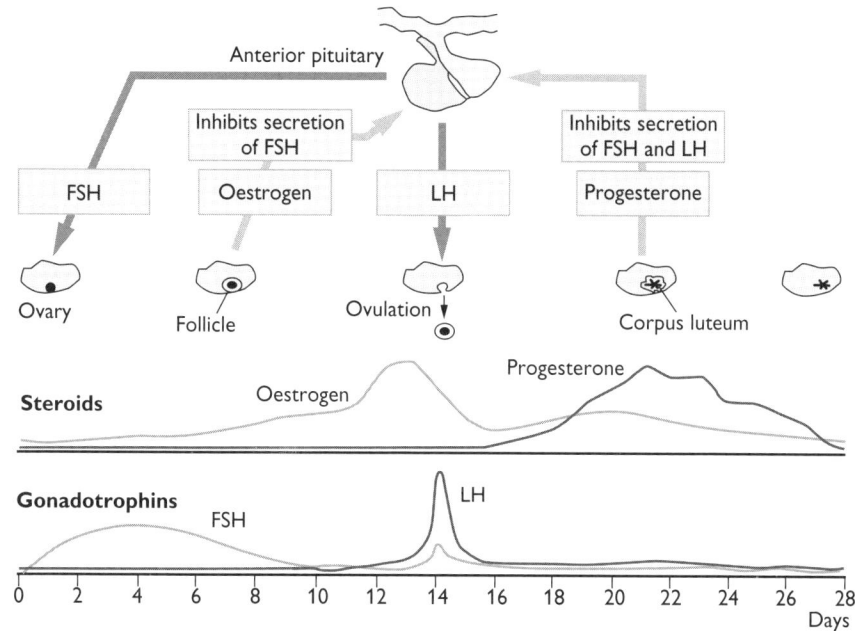

Figure 48 *The changes in the concentrations of the hormones controlling the human menstrual cycle*

The changes in plasma concentrations of all these hormones are controlled by negative feedback systems. As soon as the level of a hormone rises sufficiently, it initiates a mechanism that reduces its concentration.

However, there are also positive feedback loops that operate to control the secretion of oestrogen and progesterone. An increase in the concentration of each of these hormones in the plasma stimulates cells in the follicle to further increase their secretion of the hormone. This has the benefit that concentrations of the steroid hormones can be increased without the need for increased levels of FSH and LH. If these were too high for too long, multiple ovulation could result.

What the examiners could ask you to do

- Explain any of the key concepts.
- Recall and show understanding of any of the key facts.
- Interpret graphs showing changes in the concentration of plasma glucose, body temperature and reproductive hormones.
- Use your knowledge of the properties of enzymes to explain the importance of homeostasis.
- Use your knowledge of the relationship between size and surface area-to-volume ratio to explain why large reptiles are better able to maintain a near constant core temperature than smaller ones.
- Use your knowledge of the structure of arterioles and capillaries to explain their roles in thermoregulation.

- Use your knowledge of transmission across synapses and the oxygen dissociation curve of haemoglobin to explain some of the events in muscle contraction.
- Use your knowledge of the structure and properties of glycogen and glucose to explain the significance of their interconversion in the body.
- Use your knowledge of the tertiary structure of protein molecules to explain how hormones target specific cells.
- Use your knowledge of water potential to explain some of the effects and symptoms of diabetes.
- Comment on the reliability and/or validity of investigations into homeostatic mechanisms, bearing in mind:
  - the numbers of organisms or people used and the number of repeats carried out
  - possible flaws in the technique or in the apparatus
  - the ways in which changes were measured
  - overlap of error bars in graphs of results
  - the extent to which appropriate controls were carried out

The genetic code, protein synthesis and gene mutation

A brief revision of the structure of DNA

The DNA molecule is made from two strands arranged into a **double helix:**
- Each strand is a polynucleotide consisting of four types of **nucleotide**.
- The two strands are held together by hydrogen bonds.
- The two strands are **anti-parallel**.
- Only one strand of DNA holds coding information; this is the **coding (sense) strand**; the other strand is the **non-coding (antisense) strand**.
- Each nucleotide contains the pentose sugar **deoxyribose**, a **phosphate group** and an **organic (nitrogenous) base**.
- Each polynucleotide strand is held together by the '**sugar–phosphate backbone**' in which condensation links form between the phosphate group of one nucleotide and the deoxyribose sugar of the next nucleotide.
- The four organic bases are **adenine, thymine, cytosine** and **guanine**.
- Specific base pairing occurs between the strands:
  - adenine is always paired with thymine
  - cytosine is always paired with guanine
- In addition to the coding regions (**exons**) within a gene, there are non-coding sections of DNA called **introns**, which separate the exons.

- There are also base sequences between genes that do not code for amino acids and which repeat themselves; these are **minisatellites** and **microsatellites**.

The structure of DNA is shown in Figure 49.

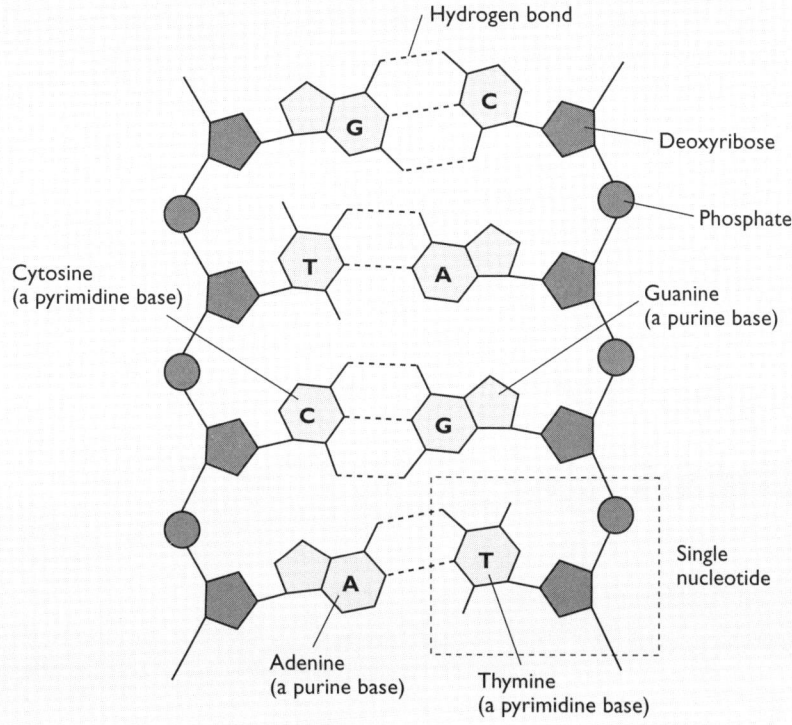

Figure 49 The structure of DNA

A brief revision of the structure of proteins

Protein molecules are **macromolecules**; they are **polymers** of amino acids. The amino acids are joined together by **peptide bonds** to form a **polypeptide chain**. Each protein molecule has three levels of organisation:
- **primary structure**
- **secondary structure** (either an α-helix or β-pleated sheet)
- **tertiary structure**

These levels of organisation are described in the table below and shown in Figure 50.

A2 Biology

Level	Bonds holding structure in place	Description	Notes
Primary	Peptide bonds	Sequence of amino acids in the polypeptide chain	Determined by sequence of triplets of bases in DNA
Secondary	Hydrogen bonds	α-helix or β-pleated sheet	Formed by folding polypeptide chain; both types can exist in the same protein molecule
Tertiary	Ionic, hydrogen and disulphide bridges	Globular of fibrous structure	Gives molecule unique shape and specific function

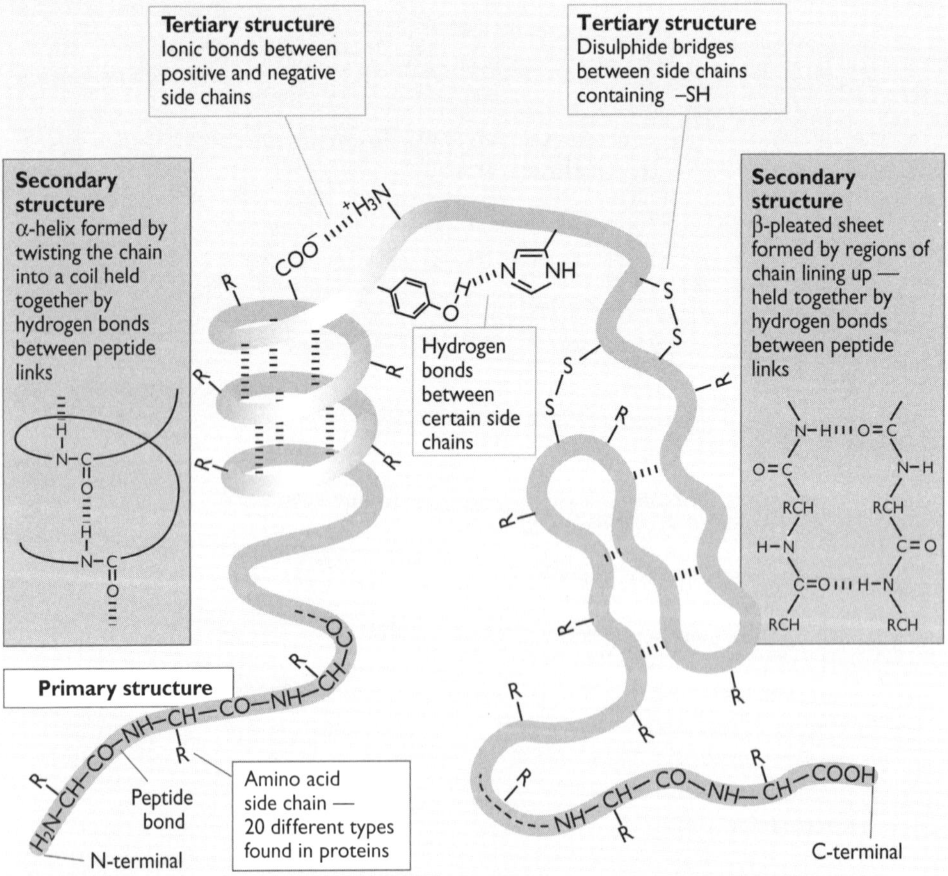

Tertiary structure
Ionic bonds between positive and negative side chains

Tertiary structure
Disulphide bridges between side chains containing –SH

Secondary structure
α-helix formed by twisting the chain into a coil held together by hydrogen bonds between peptide links

Secondary structure
β-pleated sheet formed by regions of chain lining up — held together by hydrogen bonds between peptide links

Hydrogen bonds between certain side chains

Primary structure

Peptide bond

N-terminal

Amino acid side chain — 20 different types found in proteins

C-terminal

Figure 50 The primary, secondary and tertiary structures of a protein

The genetic code and protein synthesis

Key concepts you must understand

A gene is a region in the coding strand of a DNA molecule that codes for a particular protein. The code for the protein is determined by the sequence of organic bases within the gene.

The 'code' has to be carried to the ribosomes, where protein synthesis takes place. The sequence of events (Figure 51) is as follows:
- Because of its size, DNA cannot move out of the nucleus, so the DNA code is rewritten in a molecule of **messenger RNA** (mRNA); this rewriting of the code is called **transcription**.
- The mRNA travels from the nucleus through pores in the nuclear envelope to the ribosomes.
- Free amino acids are carried from the cytoplasm to the ribosomes by molecules of **transfer RNA** (tRNA).
- The ribosome 'reads' the mRNA code and assembles the amino acids into a protein; this is called **translation**.

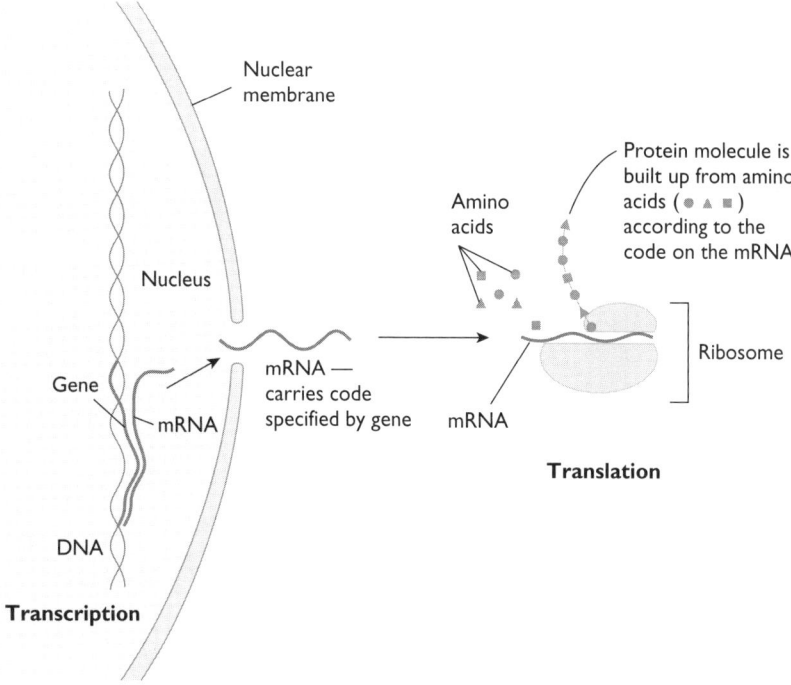

Figure 51 An overview of protein synthesis

In the transcription of DNA to mRNA the non-coding (antisense) strand is used as the template to synthesise the mRNA. The mRNA will have a base sequence that is

complementary to the antisense strand of the DNA and, therefore, the *same* as the sense strand (except that uracil replaces thymine).

The genetic code is:
- A triplet code — three bases code for one amino acid. In DNA, the three bases coding for one amino acid make up a **triplet**. In mRNA, the three bases coding for one amino acid make up a **codon**.
- Degenerate — there are more 'codes' than are needed. There are 20 amino acids and 64 triplets, so some amino acids have more than one code and some DNA triplets say 'stop transcribing' (some RNA triplets say 'stop translating').
- Non-overlapping — the three bases that form one triplet are not part of any other triplets.
- Universal — the codes are the same in all organisms.

The tables show the DNA triplets and mRNA codons that code for the 20 amino acids used in protein synthesis. They also show the stop codes. You need not learn any of these, but you may have to convert a DNA code for a particular amino acid into the corresponding mRNA code (or vice versa).

The DNA triplets and the amino acids they code for are shown in the table below.

First position	Second position				Third position
	T	C	A	G	
T	TTT ⎤ Phe TTC ⎦ TTA ⎤ Leu TTG ⎦	TCT ⎤ TCC ⎥ Ser TCA ⎥ TCG ⎦	TAT ⎤ Tyr TAC ⎦ TAA stop TAG stop	TGT ⎤ Cys TGC ⎦ TGA stop TGG Trp	T C A G
C	CTT ⎤ CTC ⎥ Leu CTA ⎥ CTG ⎦	CCT ⎤ CCC ⎥ Pro CCA ⎥ CCG ⎦	CAT ⎤ His CAC ⎦ CAA ⎤ Gln CAG ⎦	CGT ⎤ CGC ⎥ Arg CGA ⎥ CGG ⎦	T C A G
A	ATT ⎤ Ile ATC ⎥ ATA ⎦ ATG Met	ACT ⎤ ACC ⎥ Thr ACA ⎥ ACG ⎦	AAT ⎤ Asn AAC ⎦ AAA ⎤ Lys AAG ⎦	AGT ⎤ Ser AGC ⎦ AGA ⎤ Arg AGG ⎦	T C A G
G	GTT ⎤ GTC ⎥ Val GTA ⎥ GTG ⎦	GCT ⎤ GCC ⎥ Ala GCA ⎥ GCG ⎦	GAT ⎤ Asp GAC ⎦ GAA ⎤ Glu GAG ⎦	GGT ⎤ GGC ⎥ Gly GGA ⎥ GGG ⎦	T C A G

The mRNA codons and the amino acids they code for are shown in the table below.

		Second position				
		U	C	A	G	
First position (5' end)	U	UUU ⎤ Phe UUC ⎦ UUA ⎤ Leu UUG ⎦	UCU ⎤ UCC ⎥ Ser UCA ⎥ UCG ⎦	UAU ⎤ Tyr UAC ⎦ UAA stop UAG stop	UGU ⎤ Cys UGC ⎦ UGA stop UGG Trp	U C A G	Third position (3' end)
	C	CUU ⎤ CUC ⎥ Leu CUA ⎥ CUG ⎦	CCU ⎤ CCC ⎥ Pro CCA ⎥ CCG ⎦	CAU ⎤ His CAC ⎦ CAA ⎤ Gln CAG ⎦	CGU ⎤ CGC ⎥ Arg CGA ⎥ CGG ⎦	U C A G	
	A	AUU ⎤ AUC ⎥ Ile AUA ⎦ AUG Met	ACU ⎤ ACC ⎥ Thr ACA ⎥ ACG ⎦	AAU ⎤ Asn AAC ⎦ AAA ⎤ Lys AAG ⎦	AGU ⎤ Ser AGC ⎦ AGA ⎤ Arg AGG ⎦	U C A G	
	G	GUU ⎤ GUC ⎥ Val GUA ⎥ GUG ⎦	GCU ⎤ GCC ⎥ Ala GCA ⎥ GCG ⎦	GAU ⎤ Asp GAC ⎦ GAA ⎤ Glu GAG ⎦	GGU ⎤ GGC ⎥ Gly GGA ⎥ GGG ⎦	U C A G	

Key facts you must know and understand

Transcription of DNA to messenger RNA

When compared with DNA, a molecule of mRNA is:
- smaller
- single stranded (not double stranded)
- contains the pentose sugar ribose (not deoxyribose)
- contains the base uracil in place of thymine

Transcription (Figure 52) takes place in the following way.
- The enzyme DNA-dependent RNA polymerase (RNA polymerase) binds with a section of DNA next to the gene that is to be transcribed
- Transcription factors (see page 62) activate the enzyme
- RNA polymerase begins to 'unwind' the strands of DNA in the region that makes up the gene and moves along the antisense strand, using it as a template for synthesising mRNA.
- RNA polymerase assembles free RNA nucleotides into a chain in which the base sequence is complementary to the base sequence on the antisense strand of the DNA.
- The completed molecule leaves the DNA; the strands of DNA rejoin and re-coil.

The mRNA molecule now contains transcripts of:
- the exons (regions that code for amino acids)
- the introns (non-coding regions that separate the coding regions)

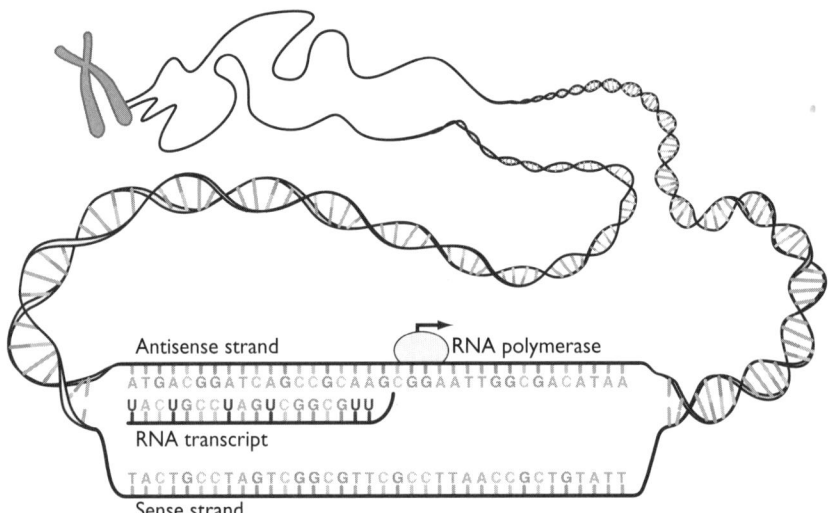

Figure 52 Transcription — the base sequence of the mRNA is identical to the sense strand of DNA, except that T is replaced by U

At this stage, the transcribed molecule is referred to as **pre-mRNA**.

Next, the introns are 'cut out' and the remaining exons are joined or spliced together. The triplets of bases in mRNA are called **codons**.

The production of functional mRNA from pre-RNA is shown in Figure 53.

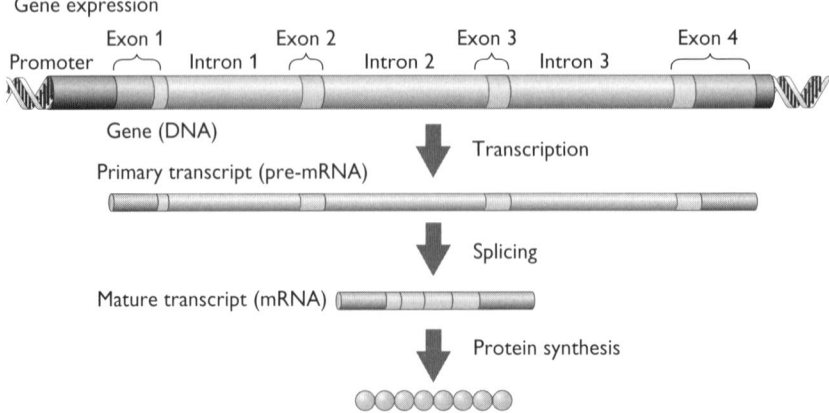

Figure 53 Producing functional mRNA from pre-mRNA

Translation of messenger RNA to a polypetptide
The structure of transfer RNA (tRNA)
Each transfer RNA molecule carries an amino acid to a ribosome. Each of the different types of tRNA has a structure that allows it to transfer just one specific amino acid *and* to be recognised as carrying that amino acid (and no other).

However, all tRNA molecules have the same basic structure (Figure 54), with two key features:
- At one end of the molecule there is a triplet of bases called an **anticodon** with a base sequence complementary to one of the mRNA codons.
- At the other end of the tRNA molecule there is an **attachment site** for the amino acid specified by the mRNA codon.

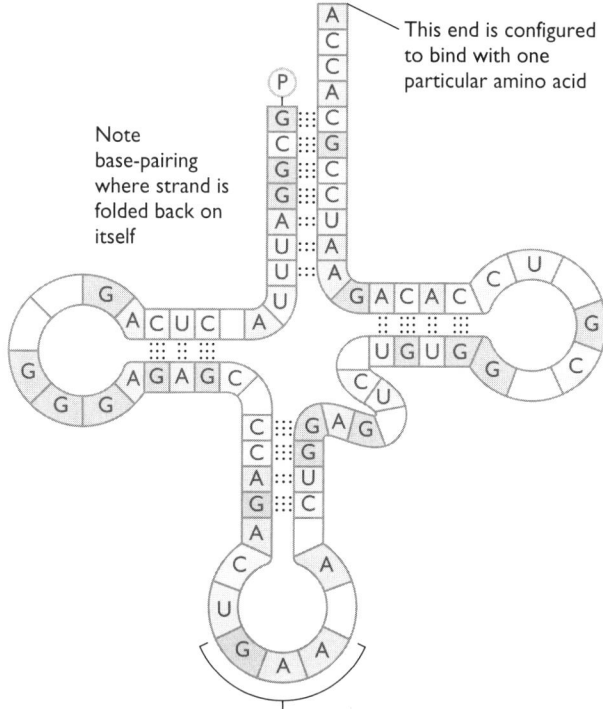

The anticodon is complementary to the mRNA codon that specifies the amino acid carried by this tRNA

Figure 54 The structure of tRNA

The stages in translation of mRNA into a polypeptide
First, mRNA arrives at a ribosome and is recognised by the ribosome. The sequence of events is then as follows:
- The first two codons of the mRNA enter the ribosome.
- Transfer RNA molecules (with amino acids attached), that have complementary anti-codons to the first two codons of the mRNA, enter the ribosome and bind to them.
- A peptide bond forms between the amino acids carried by the two tRNA molecules.
- The ribosome moves along the mRNA by one codon, bringing the third codon into the ribosome.
- The tRNA that was bound to the first codon is freed and returns to the cytoplasm.
- A tRNA with a complementary anticodon binds with the third codon, bringing its amino acid into position next to the amino acid on the second tRNA.
- A peptide bond forms between the second and third amino acids.

- The ribosome moves along the mRNA by one codon and the process is repeated until a stop codon is in position and translation ceases.

Translation is shown in Figure 55.

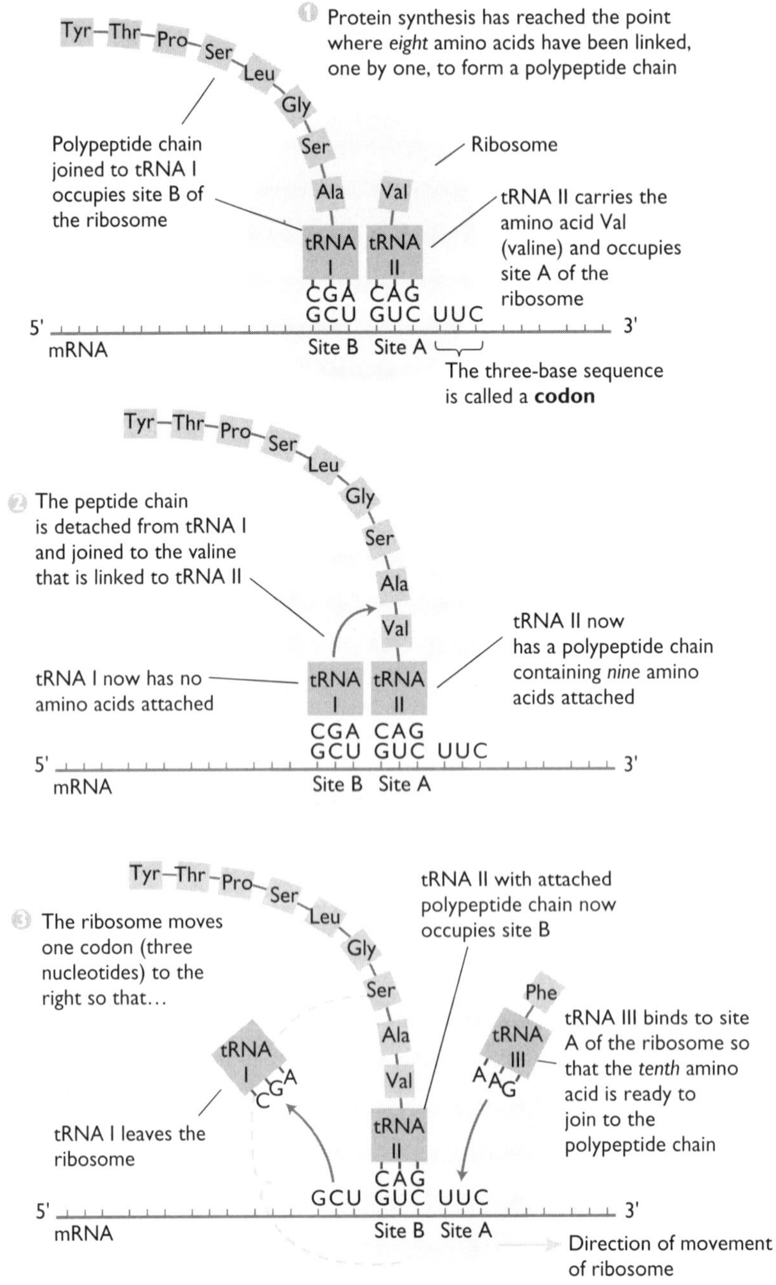

① Protein synthesis has reached the point where *eight* amino acids have been linked, one by one, to form a polypeptide chain

Polypeptide chain joined to tRNA I occupies site B of the ribosome

Ribosome

tRNA II carries the amino acid Val (valine) and occupies site A of the ribosome

The three-base sequence is called a **codon**

② The peptide chain is detached from tRNA I and joined to the valine that is linked to tRNA II

tRNA I now has no amino acids attached

tRNA II now has a polypeptide chain containing *nine* amino acids attached

③ The ribosome moves one codon (three nucleotides) to the right so that…

tRNA I leaves the ribosome

tRNA II with attached polypeptide chain now occupies site B

tRNA III binds to site A of the ribosome so that the *tenth* amino acid is ready to join to the polypeptide chain

Direction of movement of ribosome

Figure 55 The stages in translation of mRNA

Gene mutation

Key concepts you must understand

A mutation is any spontaneous change in the DNA molecule. You need only know about **point mutations** — these involve changes to just one DNA triplet. They occur most often when DNA is replicating. Two examples of point mutations are substitutions and deletions:

- In a **substitution**, one base in a DNA triplet is replaced by another.
- In a **deletion**, one base is missed out (not copied).

In the substitution shown in Figure 56 guanine replaces thymine:

$$\text{A-T-T -T-C-C -G-T-T -A-T-C ...}$$
↑
Original base

$$\text{A-T-G -T-C-C -G-T-T -A-T-C ...}$$
↑
Substituted base

Figure 56 Substitution

- The triplet ATT has been changed to ATG.
- No other triplet is affected (this is true of all substitutions).
- The original triplet, ATT, codes for the amino acid isoleucine; the new triplet, ATG, codes for methionine.
- The protein synthesised will contain one different amino acid to that coded for by the un-mutated gene; this may or may not be significantly different from the original.

Because the DNA code is degenerate, all substitutions do not result in a change in the protein synthesised. If ATT had mutated to ATC, the mutated triplet would still code for isoleucine.

In a deletion (Figure 57), to replace the base that is 'missed out' in the mutated DNA, the first base of each original triplet becomes the last base of the preceding triplet. This is called a **frameshift** and produces a totally new base sequence that results in a nonsense code and either a non-functional protein or no protein at all.

$$\text{A-T-T -T-C-C -G-T-T -A-T-C ...}$$
↑
Deletion here

$$\text{A-T-T -C-C-G -T-T-A -T-C ...}$$
↑
Replaced by first base of next triplet

Figure 57 Deletion

In the example in Figure 57, the first triplet remains the same because the first base in the second triplet happens to be the same as the deleted third base in the first triplet. All the triplets after the deletion are altered.

Mutations that occur in ordinary body cells are not passed on to the next generation. For this to happen, the mutation must take place in either:
- a sex cell, or
- a cell in the sex organ that divides to form the sex cells

Key facts you must know and understand

Mutations occur spontaneously and randomly. However, the rate of mutation can be increased by a number of factors including:
- carcinogenic chemicals — for example those in tobacco smoke
- high energy radiation — for example ultraviolet radiation, X-rays

A mutation in a normal body cell could have one of several consequences, including:
- it will be completely harmless
- it will damage the cell
- it will kill the cell
- it will make the cell cancerous, which might kill the person

Proto-oncogenes and **tumour suppressor genes** play important roles in regulating cell division and preventing the formation of a tumour. The flowchart (Figure 58) shows how mutations in these genes, if they are not repaired, can result in tumour formation.

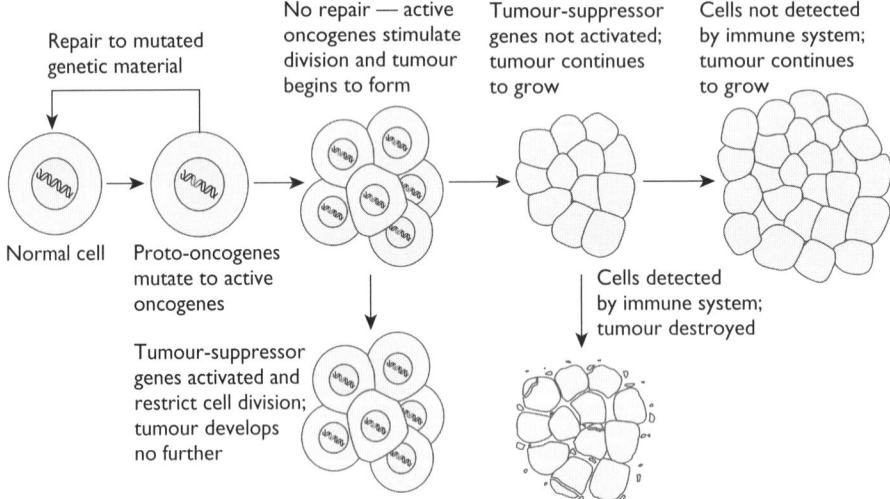

Figure 58 How mutations may result in tumour formation

What the examiners could ask you to do

- Explain any of the key concepts.
- Recall and show understanding of any of the key facts.
- Interpret diagrams of protein synthesis.
- Convert DNA codes into RNA codes and vice versa.

- Use your knowledge of complementary base pairing to explain transcription and translation
- Use your knowledge of protein structure to explain why frameshift mutations result in the synthesis of non-functional proteins

The control of gene action

Key concepts you must understand

In most cells, only a small fraction of the genes are active. The active genes vary between different types of cell and also at different times.

Not all DNA codes for proteins. Non-coding DNA includes:
- introns within genes
- minisatellite and microsatellite regions between genes
- **promoter** and **enhancer** regions of DNA close to coding DNA that RNA polymerase binds with to initiate transcription of that gene

Promoter (and enhancer) regions of DNA also need to be switched 'on' and 'off' otherwise the gene regulated would be permanently active or permanently inactive.

Genes can be actively 'switched on' in the following way. Proteins called **transcription factors** that are present in the cell bind with the promoter and enhancer regions. RNA polymerase is then able to recognise the 'promoter/transcription factor complex' and transcribe the gene (Figure 59).

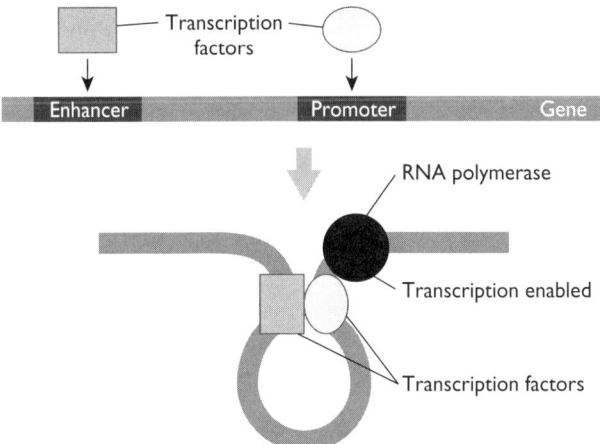

Figure 59 Transcription factors allow RNA polymerase to bind to and transcribe a gene

Genes can also be actively 'switched off'. This happens when a **repressor** molecule binds with the promoter region and prevents transcription factors from binding. In this state, RNA polymerase cannot bind and transcribe the gene.

Genes can also be '**silenced**'. This is not the same as repressing the action of the gene as the gene is still active. Cells can 'silence' genes by degrading the mRNA that is transcribed from the gene. So, although the gene is still active, no protein is synthesised because no mRNA reaches the ribosomes to allow translation to occur. Gene silencing by degrading mRNA is one example of **post-transcriptional repression**.

The reason why different cells express different genes is because they contain different transcription factors and different repressors.

Some cells have the potential to express all their genes. These cells are called **totipotent cells**. Properties of totipotent cells include:
- They are unspecialised (they are not specialised for any particular function as, say, a neurone or rod cell is).
- They are capable of dividing and renewing themselves for long periods.
- They are capable of giving rise to specialised cells.

Specialisation of cells during development occurs by genes being repressed. The only genes that are capable of being expressed in a neurone are those that are needed for that particular function. The gene for brown pigmentation of the iris (for example) has been repressed.

Unspecialised cells that divide to give specialised cells are called **stem cells**. There is a hierarchy of specialisation. The hierarchy in humans is shown in Figure 60.

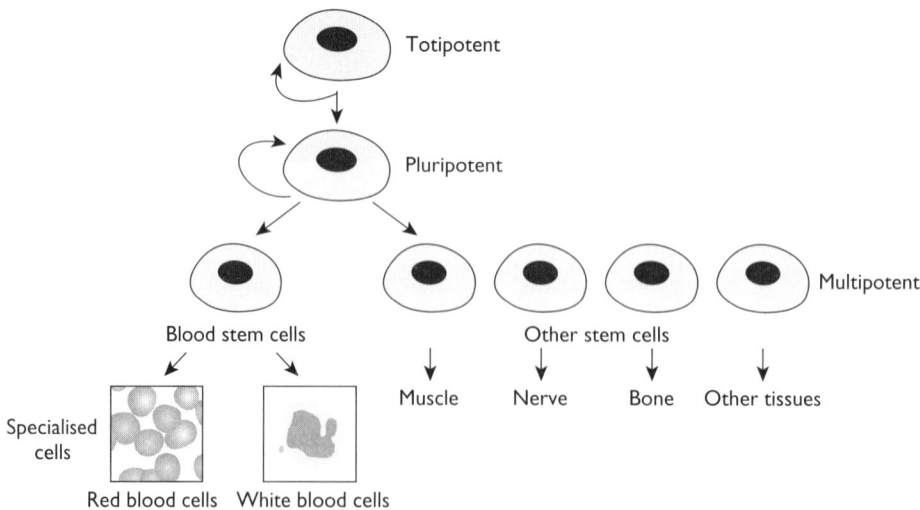

Figure 60 The hierarchy of cell specialisation in humans

When they divide:
- totipotent cells can give rise to any kind of tissue in the body and to the cells of the placenta
- pluripotent cells can give rise to any kind of tissue but not the cells of the placenta
- multipotent cells can give rise to specialised cells

Because totipotent and multipotent stem cells can give rise to any type of cell in the adult body, they have the potential to be used to treat medical conditions in which cells are degenerating or have been damaged. The stem cells divide and form the type of cell that is present in the tissue into which they are placed. These new cells replace the cells that are being lost or that have been damaged.

Key facts you must know and understand

Transcription factors
One model of the way in which transcription factors regulate gene action is as follows:
- The transcription factor binds to a **promoter sequence** of DNA near to the gene to be activated.
- RNA polymerase binds to the promoter sequence–transcription factor complex.
- RNA polymerase is 'activated' and moves away from the DNA–transcription factor complex along the DNA of the coding gene.
- The RNA polymerase transcribes the antisense strand of the DNA as it moves along; the gene is now being expressed.

The female sex hormone oestrogen binds with receptors in certain cells to form an oestrogen–receptor complex. This complex acts as a transcription factor and binds to promoter regions of genes that stimulate cell division. This allows the hormone to produce its effects of:
- causing cells in the uterus lining to divide
- causing cells in the breasts to divide

Gene silencing
Gene silencing (Figure 61) sometimes occurs as a result of the action of **short interfering RNA** (**siRNA**). These RNA molecules are unusual because they are:
- very short — only about 21–23 nucleotides long
- double stranded

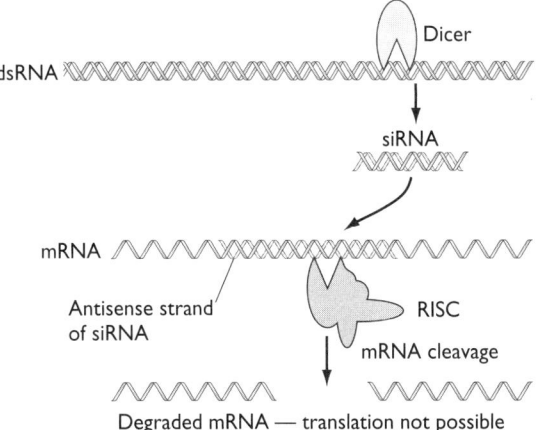

Figure 61 Gene silencing by siRNA

Short interfering RNA molecules act in the following way:
- siRNA is produced in the nucleus from a range of genes.
- The antisense strand of the siRNA then binds with the mRNA it is to silence.
- This guides a complex of molecules called RISC to the site.
- Together with the siRNA, RISC breaks down (degrades) the mRNA.

Totipotent cells in plants

More cells in plants are either totipotent or pluripotent than is the case in animals. Because of this, mature plants can be grown from small parts of plants called **cuttings**. One way to take a cutting is as follows:
- Cut a small side shoot from a plant.
- Remove some leaves (to prevent excessive water loss).
- Place the cutting in compost and keep watered.

The cutting develops small roots at the base and, eventually, a full root system. This means that the cells at the base of the cutting must have started to divide again and the cells they formed then specialised into the various tissues in the roots. To do this, they must have been at least pluripotent.

In micropropagation, the totipotency or pluripotency of some plant cells is exploited even further. Small sections of plant tissue called **explants** are grown in special culture media. The explants are treated first with hormones that make them develop roots and later with hormones that make them develop shoots. When they have developed sufficiently, the young plantlets are transplanted into compost and grown in glasshouses.

What the examiners could ask you to do

- Explain any of the key concepts.
- Recall and show understanding of any of the key facts.
- Interpret diagrams and flow charts of the control of gene action.
- Use your knowledge of protein synthesis to explain how gene silencing could potentially treat conditions caused by abnormal protein synthesis.
- Comment on the ethics and validity of findings from research into the use of stem cell treatments, bearing in mind:
  - the numbers of organisms or people used and the number of repeats carried out
  - the species used in the investigations
  - the balance between any harm to the organisms used and the potential gain
  - the ways in which changes were measured
  - the overlap of error bars in graphs of results
  - the extent to which appropriate controls were carried out

Gene cloning technology and its applications

Gene cloning technology

Key concepts you must understand

Cloning genes produces multiple, identical copies of a gene. There are two main types of gene cloning:
- **in vivo cloning** — the gene is introduced into a cell and is copied as the cell divides
- **in vitro cloning** — DNA (containing the gene in question) is copied many times over by **polymerase chain reaction** (**PCR**) in a PCR machine

Gene (DNA) probes can help to identify genes. A gene probe is a short length of single-stranded DNA that is radioactive (or sometimes fluorescent). The sequence of bases is complementary to at least part of the sequence of bases in the gene of interest. When a radioactive gene probe is mixed with a sample of DNA containing the gene for which it is specific, the gene probe binds and makes the DNA sample radioactive. This can be detected using X-ray film.

Once identified, the gene to be cloned can be obtained by two main methods:
- It can be extracted from a donor cell using enzyme technology.
- It can be created in vitro from the corresponding mRNA.

If insufficient copies of the gene are obtained, the amount can be increased using the PCR. This process mimics the semi-conservative replication of DNA in cells.

Once the gene has been cloned, it may be:
- stored in a **gene library**
- transferred into another cell, often from a different species, which, as a result, becomes **transgenic** (has a gene from a different species)

Vectors (carriers) are needed to transfer DNA into another cell. There are two types of vector:
- **plasmids** (small circular pieces of DNA found in bacteria)
- **viruses** that have been modified to be non-pathogenic

Gene sequencing is a technique that allows biologists to determine the sequence of bases in a section of DNA.

The first gene sequencing technique was the **chain terminator technique** developed by Fred Sanger. This technique works by:
- creating many sections of DNA complementary to part of one strand of the DNA being investigated

- ensuring that some sections are just one base pair long, some two, some three and so on all the way up to sections that are 'full length'
- finding out which base is at the end of each strand

The sequence of bases at the ends of the newly created strands is complementary to the sequence in the DNA being investigated.

Many gene sequencing techniques can only determine the base sequence of relatively small sections of DNA. To determine the base sequence of larger sections, gene sequencing is sometimes combined with **restriction mapping**. This technique uses **restriction enzymes**. These enzymes cut DNA at specific **restriction sites** — specific base sequences, for example CTTAAG. The size of the DNA fragments produced can be determined by gel electrophoresis.

Gel electrophoresis is a technique used to separate charged particles. Negatively charged ions move through a gel towards a positive electrode; positively charged ions move towards a negative electrode. The gel acts as a 'molecular sieve', slowing down larger particles which, therefore, do not move as far as smaller ones.

Using two (or more) restriction enzymes together with gel electrophoresis allows us to make predictions about possible locations of their restriction sites in the DNA.

Using gel electrophoresis to check the size of the fragments produced by treatment with both restriction enzymes simultaneously will confirm one of the predictions (see Figure 62).

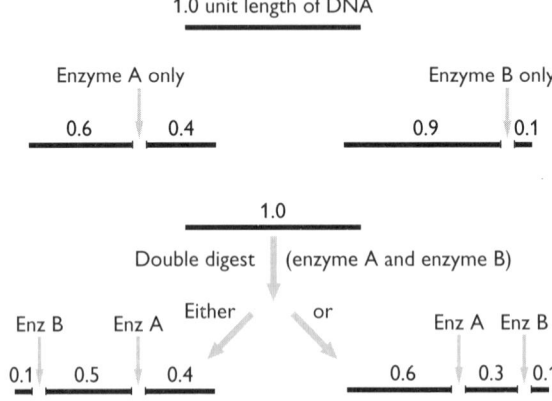

Figure 62 Predicting the location of restriction sites in DNA

To gene-sequence a really large section of DNA (Figure 63), we could:
- digest a large piece of DNA into several smaller pieces
- construct a restriction map for each piece
- compare these restriction maps and see if there was any overlap of the restriction sites, which would suggest a common piece of DNA

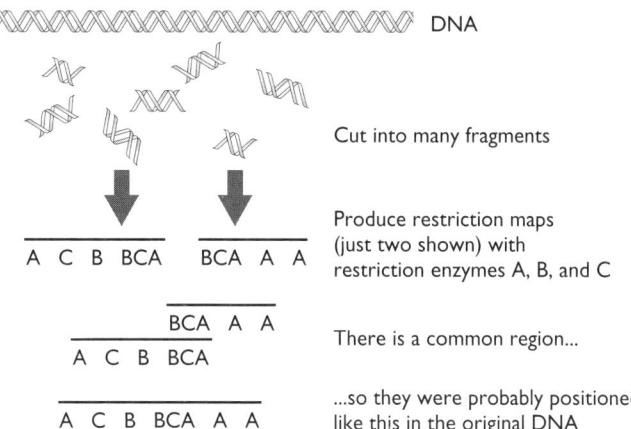

Figure 63 Using restriction mapping to determine base sequences of large fragments of DNA

Once the DNA fragments have been sequenced, we can now assemble the base sequences into a larger sequence, remembering not to duplicate the overlaps.

Key facts you must know and understand

In vivo gene cloning

The flow chart in Figure 64 shows the main stages of in vivo gene cloning.

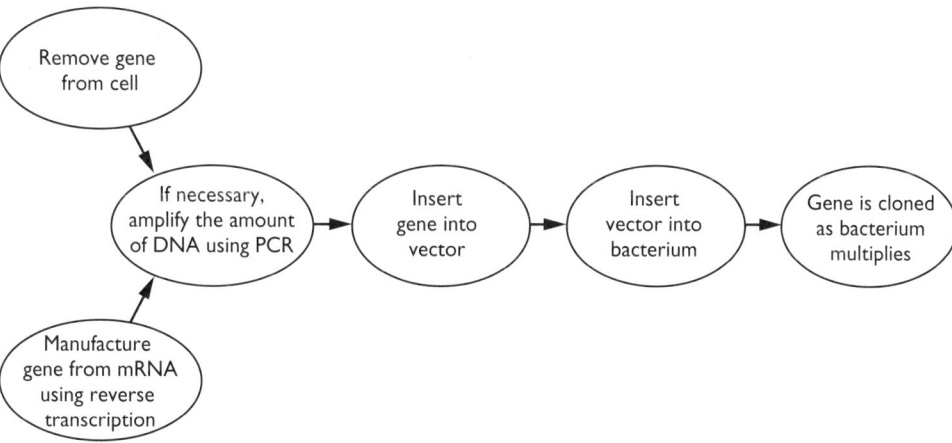

Figure 64 The main stages of in vivo gene cloning

Removing the gene from the donor cell

The donor cells are incubated with restriction enzymes whose restriction sites will ensure that a section of DNA containing the gene is isolated.

Some restriction enzymes do not make a 'clean cut' across the two strands of the DNA, but make a staggered cut, leaving unpaired bases. These staggered ends are often called '**sticky ends**' (Figure 65).

A2 Biology

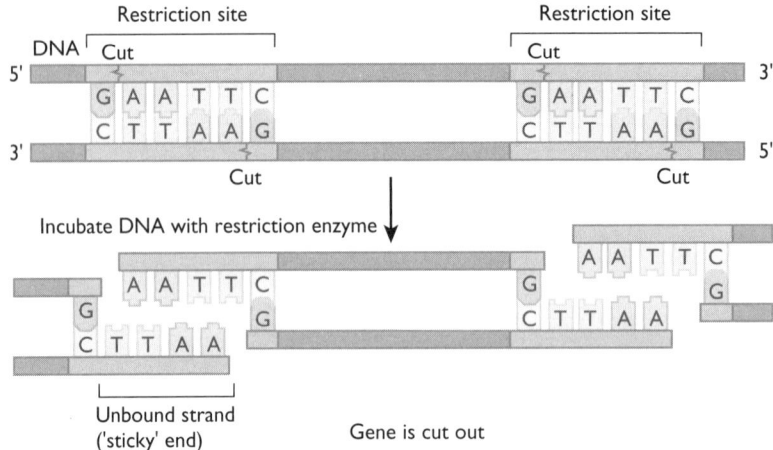

Figure 65 Sticky ends

Creating the gene from mRNA

This technique uses an enzyme called **reverse transcriptase**. This enzyme reverses the process of transcribing DNA into mRNA and transcribes mRNA into DNA.

The main stages of the process (Figure 66) are as follows:
- Incubate the mRNA molecule that corresponds to the gene with reverse transcriptase and the necessary free nucleotides.
- Reverse transcriptase creates a single strand of complementary DNA (cDNA).
- 'Wash out' the mRNA.
- Incubate the cDNA with DNA polymerase and free DNA nucleotides.
- DNA polymerase creates a complementary strand of DNA which bonds with the strand created by reverse transcriptase.

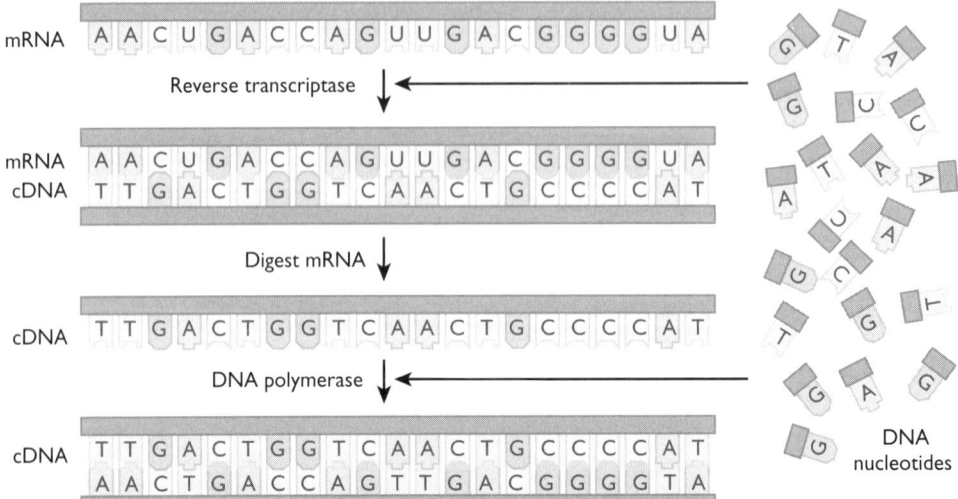

Figure 66 Creating the gene using reverse transcriptase

Amplifying the amount of DNA using PCR

The polymerase chain reaction (PCR) is an automated technique that allows a tiny sample of DNA to be amplified many times in a short period of time. There is a repeating cycle of separation of the two DNA strands, followed by synthesis of a complementary strand for each. The amount of DNA doubles with each cycle. The main stages are shown in Figure 67.

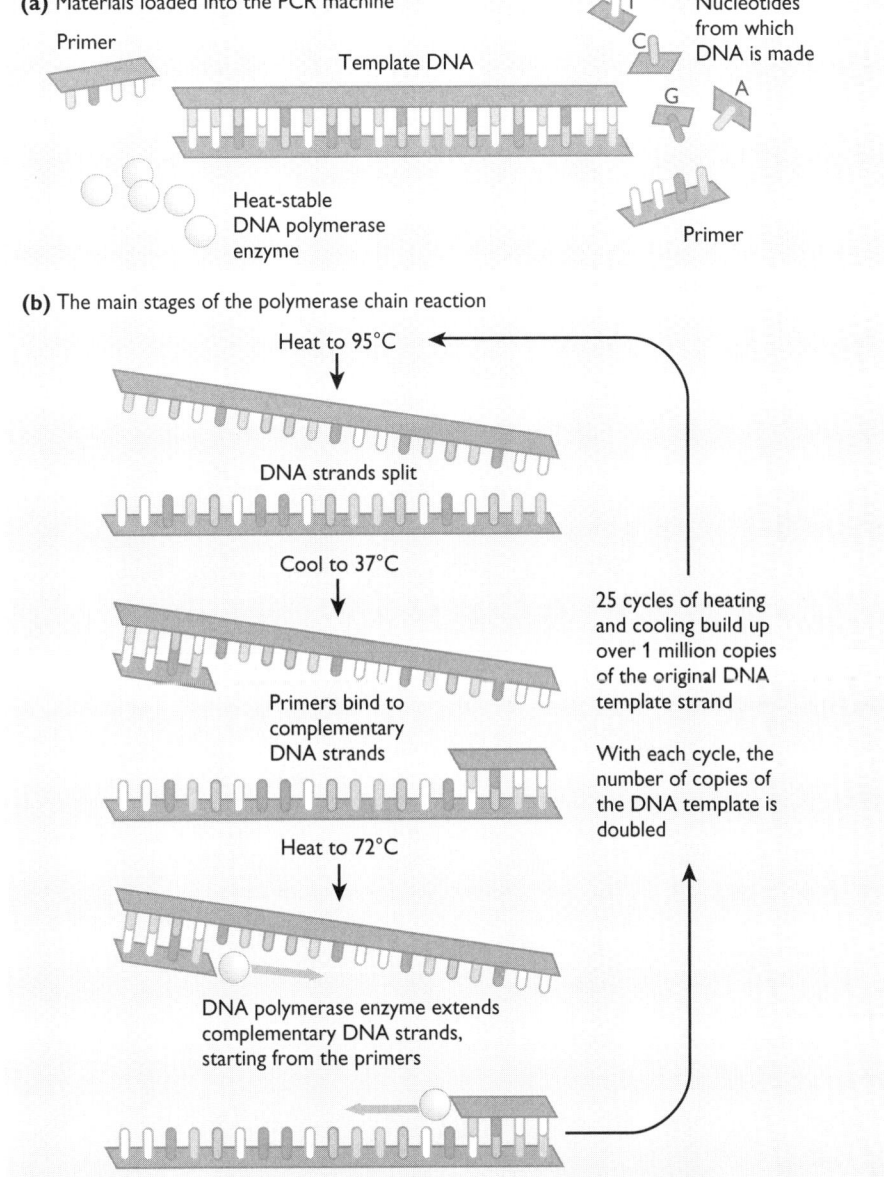

Figure 67 The polymerase chain reaction

The DNA polymerase used is thermostable so it can withstand the temperature of 95°C. It has an optimum temperature of 72°C, which is why the replication phase is carried out at that temperature.

Transferring the gene into a vector

Plasmids and viruses have both been used as vectors, but plasmids are usually preferred. The gene is transferred into a plasmid (Figure 68) as follows:
- The plasmid is cut open using the same restriction enzyme used to cut out the DNA fragments from the donor cell. The sticky ends of the two types of DNA will contain complementary base sequences (if the gene was synthesised from mRNA, sticky ends are added).
- The DNA fragments are incubated with the plasmids. The plasmid DNA and the gene DNA **anneal** (join) in the following way:
  - Hydrogen bonds form between the bases in the sticky ends, weakly holding the gene DNA in place in the plasmid.
  - Catalysed by the enzyme **ligase**, covalent bonds form between the sugar–phosphate backbones of the plasmid DNA and gene DNA; the gene has now been firmly **spliced** into the plasmid.

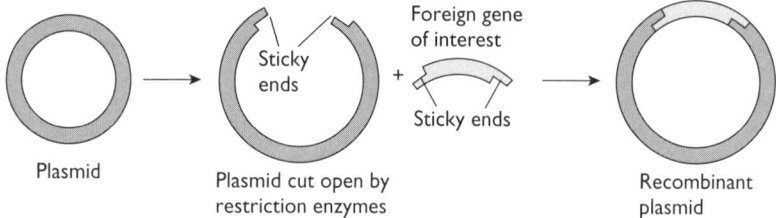

Figure 68 *Transferring a gene into a plasmid*

Any DNA that has had 'foreign DNA' inserted into it is called **recombinant DNA**, so the plasmid is now a **recombinant plasmid**.

Transferring the plasmid into a bacterium

The bacteria are treated with a solution of calcium chloride, which makes the cell wall permeable to plasmids. The bacteria are then incubated with the plasmids. However:
- the frequency of plasmid take-up by the bacteria may be as low as 1 in 10 000
- some bacteria will take up recombinant plasmids and others will take up the original, non-recombinant plasmids; those that take up the recombinant plasmids are called **transformed bacteria**

There are two main ways of checking which bacteria have taken up the recombinant plasmids.

Method 1: using marker genes

Some plasmids contain genes that confer resistance to two antibiotics — for example, both ampicillin *and* tetracycline. Introducing a gene into such a plasmid (Figure 69) splits the gene for tetracycline resistance and makes it inactive. At the end of the process, there will be the following types of bacteria:

- those that have not taken up any plasmids and are not resistant to either antibiotic
- those that have taken up the original plasmids and are resistant to both antibiotics
- those that have taken up the recombinant plasmids and are resistant to ampicillin only

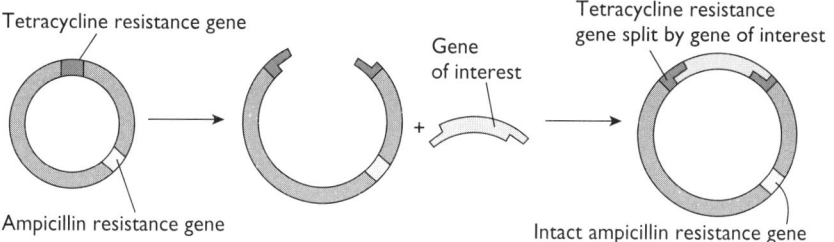

Figure 69 *Inserting a gene into a plasmid can split a gene already present*

The bacteria are cultured on media containing ampicillin or tetracycline. Those that survive on the ampicillin culture *only* are the transformed bacteria.

Method 2: using DNA probes
This technique is shown in Figure 70.

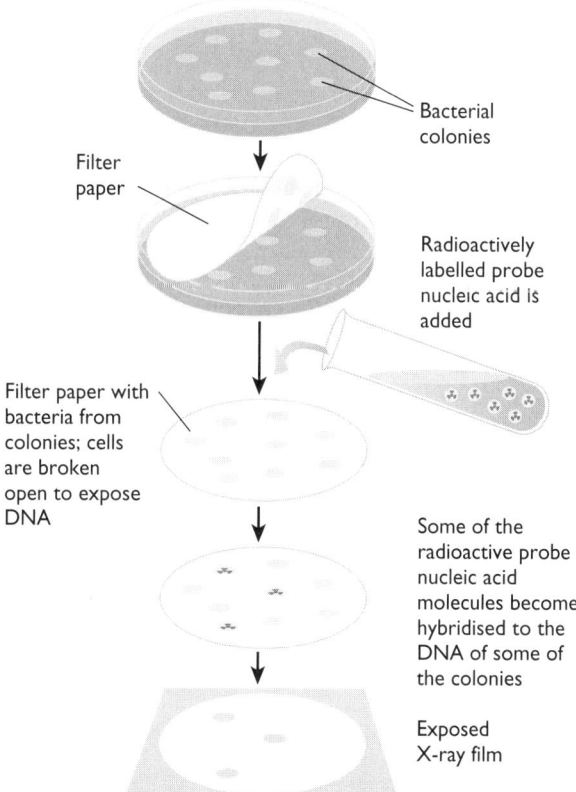

Figure 70 *Using a DNA probe to identify transformed bacteria*

- Culture the bacteria on agar gel in Petri dishes. Each bacterium will occupy a unique location on the surface of the agar.
- Incubate the Petri dish at a suitable temperature so that each bacterium multiplies to form a colony of millions (all containing copies of any plasmid taken up by the original bacterium).
- 'Blot' each Petri dish with filter paper to transfer a few cells from each colony to the filter paper in the same relative positions as the colonies in the Petri dish.
- Break open the cells on the filter paper and split the two strands of DNA.
- Incubate with a radioactive gene probe (with a sequence complementary to the gene of interest). Only DNA from cells with the gene (in the recombinant plasmid) will bind with the probe and become radioactive.
- Locate the radioactive DNA on the filter paper by producing an X-ray photograph.
- The positions of the radioactive DNA on the X-ray photograph correspond to those colonies of bacteria in the Petri dish that contain the gene of interest.

Once the bacteria that contain the gene have been identified, they can be cultured to give large numbers of bacteria for a specific purpose or simply to clone the gene.

DNA sequencing: the chain terminator technique

This technique has some parallels with PCR, as it also requires:
- a single-stranded section of DNA
- DNA polymerase
- DNA primers
- DNA nucleotides

The single-stranded DNA is primed and mixed with the polymerase, nucleotides and specially modified DNA nucleotides called dideoxynucleotides (ddNTPs). ddNTPs can form bonds with only one other nucleotide. If they enter a growing chain of DNA, the chain cannot be extended any further because the ddNTP has already bonded with its specific nucleotide.

In the chain terminator technique:
- Four samples of the single stranded DNA are primed.
- Each sample is mixed with all four normal nucleotides and one dideoxynucleotide (e.g dideoxycytosine or dideoxythymine), which is either radioactive or tagged with a fluorescent dye.
- DNA polymerase is added and it begins to build a chain complementary to the DNA sample.
- In the dideoxycytosine (ddC) reaction tube, polymerase uses ddC in the first available position (opposite guanine) on some of the strands, which then stop building. All the other strands continue building.
- In some of the others, normal cytosine is used in the first position and ddC is used in the second position. The strands then stop building.
- In other strands, the ddC is not used until the third position, in others the fourth position...and so on.
- At the end of the run in that tube, there will be new DNA strands with ddC in each available position. The ddC nucleotide will be at the end of each of these strands.

A similar process will occur in the other three reaction tubes, but with different ddNTPs at the end of the fragments.

The fragments from each tube are then separated at the same rate for the same length of time by gel electrophoresis. Figure 71 shows how strands of different lengths from the different tubes might be distributed. The different fragments move different distances, according to their masses. In this sample, the shortest fragment contained ddA and so moved furthest. The longest also contained ddA and moved least. All the strands begin with CTG because this is the region that is complementary to the primer on the DNA.

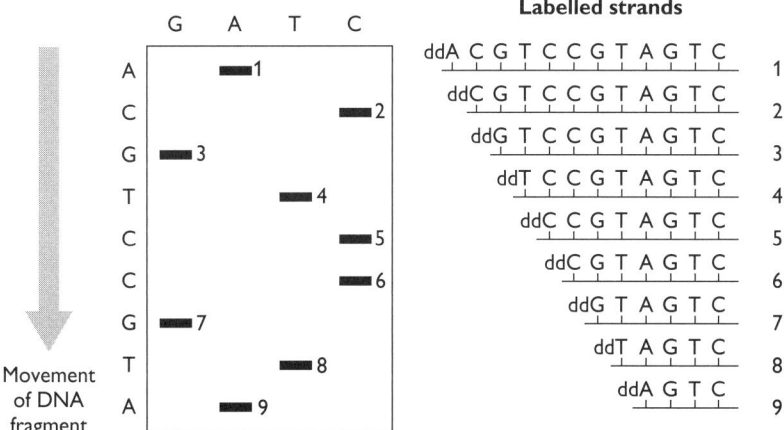

Figure 71 *Results of gel electrophoresis of DNA fragments produced by the chain terminator technique*

Applications of gene cloning technologies

Key facts you must know and understand

Genetically modified organisms are used to manufacture specific products to benefit humans (e.g. insulin, bovine somatotrophin and some vaccines). Other organisms have been genetically modified to produce increased yields (e.g. of cereals, or of milk in cattle).

Gene technology is also used to produce **DNA microarrays** (**DNA chips**) which allow the identification of, and assessment of the activity of, specific genes. A microarray contains thousands of gene probes, each specific for a particular gene. These react with a person's genes to produce a certain colour. The position of the probe in the microarray identifies the gene; the tone and the intensity of the colour allow computers to analyse the activity of the gene.

Gene sequencing and restriction mapping can identify human genes and provide information for genetic counsellors to advise patients on what options may be open to them, knowing that they have certain defective genes and may pass these genes on to their children.

Genetic fingerprinting is a technique for comparing the non-coding DNA of different people. The coding DNA within the genes does not vary a great deal between individuals — for example, the base sequence in the gene for normal haemoglobin is the same in everyone. Non-coding DNA is inherited along with the coding DNA. Genetic fingerprints can, therefore, be used to help resolve disputed parentage — each fragment of DNA in the genetic fingerprint of a child must have come from one parent or the other. Genetic fingerprints are prepared as shown in Figure 72.

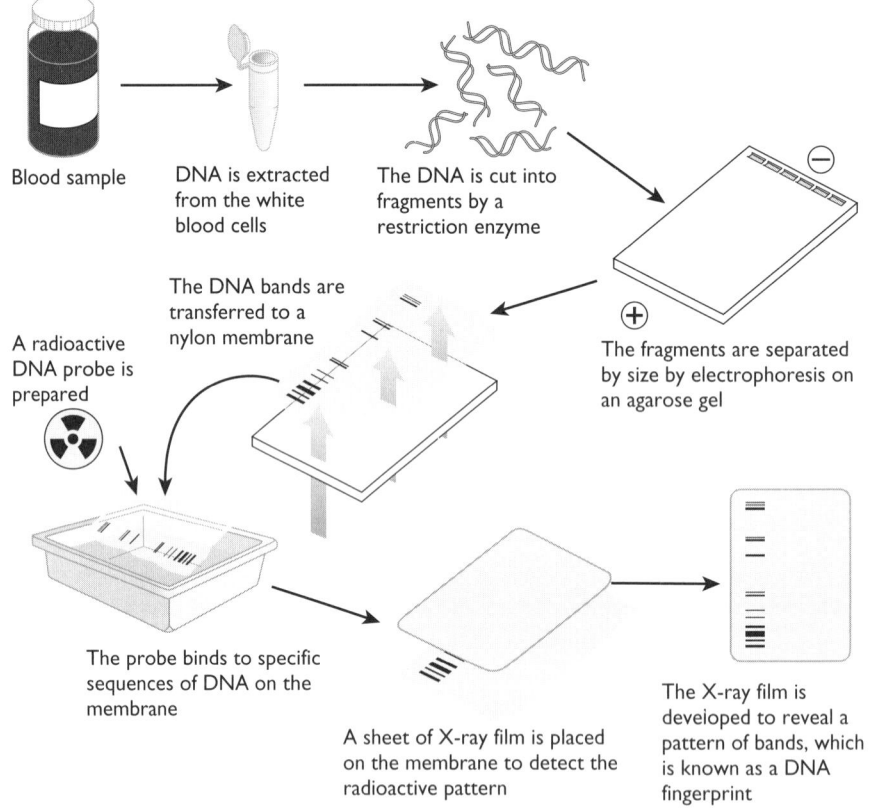

Figure 72 The main stages in obtaining a genetic fingerprint

Gene therapy involves treating genes that cause disease — for example in cystic fibrosis and in sickle-cell anaemia. Because these genes act only in some cells of the body, the cells in which they act can be targeted. Researchers in this area use three principal techniques:
- add a normal gene to the affected cells to restore normal functioning
- repair the abonormal gene by selective reverse mutation

- regulate the activity of the gene (it could be 'silenced' or turned on as required)

Much of the research to date has centred on the first technique, often using viruses as vectors for the gene. Progress has been limited because of:
- the low rate of gene take-up by cells with the defective gene
- attack by the immune system
- the virus vectors regaining virulence and causing disease

There are **moral** and **ethical** concerns about the use of gene technology:
- Morality is our personal sense of what is right and what is wrong.
- Ethics represent the 'code' adopted by a particular group to govern its way of life.

The concerns that some people have about gene technology include:
- *A species is sacrosanct and should not be altered genetically in any way.*
  People who take this moral stance usually do so on the basis that the genes from one species would not normally find their way into another species. However, genes have been 'jumping' from one species to another for millions of years.

- *Not enough is known about the long-term ecological effects of introducing genetically modified organisms into the field. They may out-compete wild plants and take over an area.*
  The effects of any new crop cannot be determined without field trials. Does this make it wrong?

- *If plants are genetically engineered to be resistant to herbicides, the gene could 'jump' into populations of weeds and other wild plants.*
  Non-genetically modified herbicide-resistant strains of many plants already exist. The gene could just as easily jump from these.

- *Gene technology might give doctors the ability to create designer babies.*
  Most doctors would find this morally and ethically unacceptable. They might consider replacing genes that cause disease but not replacing genes merely to improve a child's image in the eyes of its parents. However, if such practices become possible, who will define for doctors what is ethically acceptable? What will be the dividing line between cosmetic gene therapy and medical gene therapy?

- *Using **genetic fingerprinting** to combat crime will only be useful if there is a genetic database — a file of the genetic fingerprints of everyone in the country.*
  Who will have access to this information? If insurance companies had access to the genetic database, they might refuse insurance (or charge higher premiums) to people with an increased risk of, say, heart disease. Employers could (covertly) refuse employment to people because their 'genetic profiles' did not meet particular requirements. A recent ruling from the European Court of Human Rights states that the police have no right to hold the DNA of someone unless that person has been convicted of a crime.

What the examiners could ask you to do

- Explain any of the key concepts.
- Recall and show understanding of any of the key facts.
- Interpret diagrams and flow charts of processes.
- Discuss the ethics of applications of gene cloning technologies, from a non-personal viewpoint, balancing concerns against potential benefits from the process.
- Use your knowledge of DNA replication to explain:
  - the way in which chain terminator techniques of gene sequencing work
  - the basis of the PCR
- Comment on the ethics and validity of findings from research into gene cloning technologies and their applications, bearing in mind:
  - the numbers of organisms or people used and the number of repeats carried out
  - the species used in the investigations
  - the ways in which changes were measured
  - overlap of error bars in graphs of results
  - the extent to which appropriate controls were carried out
  - the balance between harm to the environment or organisms used and the potential gain

Questions & Answers

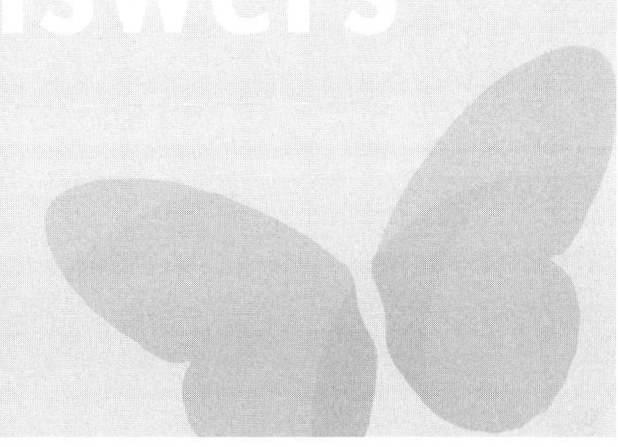

This section contains questions similar in style to those you can expect to see in your Unit 5 examination. The limited number of questions means that it is impossible to cover all the topics and all the question styles, but they should give you a flavour of what to expect. The responses that are shown are real students' answers to the questions.

There are several ways of using this section. You could:
- 'hide' the answers to each question and try the question yourself. It needn't be a memory test — use your notes to see if you can actually make all the points you ought to make
- check your answers against the candidates' responses and make an estimate of the likely standard of your response to each question
- check your answers against the examiner's comments to see if you can appreciate where you might have lost marks
- check your answers against the terms used in the question — did you *explain* when you were asked to, or did you merely *describe*?

Examiner's comments

All candidate responses are followed by examiner's comments. These are preceded by the icon 🅔 and indicate where credit is due. In the weaker answers, they also point out areas for improvement, specific problems and common errors such as lack of clarity, weak or non-existent development, irrelevance, misinterpretation of the question and mistaken meanings of terms.

Question 1

Hormone action

The level of plasma glucose at rest is maintained largely by the hormones insulin and glucagon. These are both secreted by islet cells in the pancreas and affect mainly skeletal muscle cells and liver cells.

(a) Explain why the hormones affect mainly skeletal muscle cells and liver cells. (2 marks)

(b) Explain how one molecule of glucagon can bring about the conversion of many molecules of glycogen to glucose. (2 marks)

(c) Name one other hormone that can influence the level of plasma glucose. (1 mark)

Total: 5 marks

■ ■ ■

Candidates' answers to Question 1

Candidate A
(a) Liver and skeletal muscle cells are the target cells for insulin and glucagon so they bind to these cells in particular.

Candidate B
(a) Insulin and glucagon bind to protein receptors in the plasma membranes of these cells. The protein receptors are shaped so that only these hormones will fit — other cells don't have the same receptors.

> Candidate A uses the idea of binding but does not say to what, and does not explain why these cells are the target cells. Candidate A fails to score. Candidate B makes both points, for 2 marks. To answer this synoptic question, you need an appreciation of shape of protein receptor molecules from Unit 1.

Candidate A
(b) The hormone causes the cell to break down many molecules of glycogen, one after the other.

Candidate B
(b) When one molecule of glucagon binds, it activates an enzyme which catalyses the conversion of glycogen to glucose. An enzyme can catalyse the breakdown of many molecules of glycogen. As soon as one leaves the active site of the enzyme, another one can enter to be broken down. This is the turnover rate of the enzyme.

> Candidate A does not present any information that is not given in the question, and so fails to score. You must beware of simply re-wording the question. Candidate B understands the cascade principle of hormone action, and gains 2 marks.

Candidate A

(c) Adrenaline

Candidate B

(c) Thyroxine increases the basal metabolic rate (BMR), increasing the rate at which glucose is used up.

- Both candidates gain 1 mark.

- **This question is not just about the effects of hormones. It is about how they bring about their effects and, as such, requires understanding of several areas of the specification. Candidate A should have been able to do better than 1 mark: the first section concerning targeting specific cells is straightforward if you understand the concept of specific receptor proteins. Candidate B scores all 5 marks.**

Question 2

Action potentials

Nerve impulses are propagated along the axons of neurones as a series of action potentials.

Figure 1 shows the changes in membrane potential, sodium ion (Na^+) conductance and potassium ion (K^+) conductance of an axon membrane as an action potential is generated.

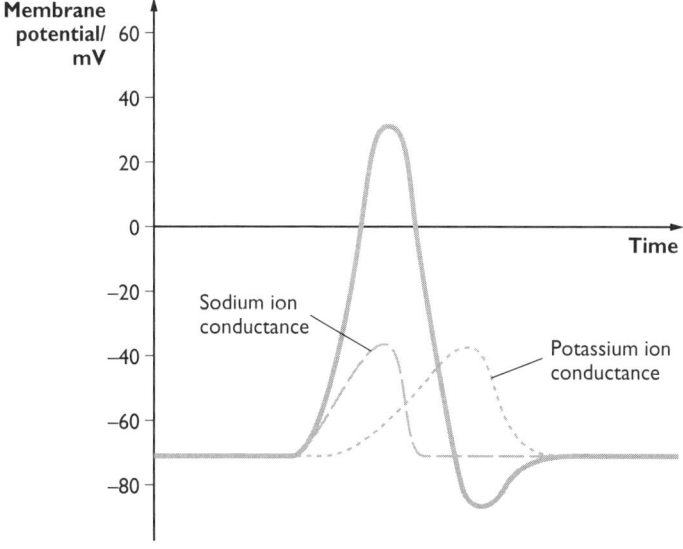

Figure 1

(a) Describe the evidence in the graphs that suggests that depolarisation is caused by an influx of sodium ions, while repolarisation is caused largely by the exit of potassium ions. (3 marks)

(b) Explain the role of the refractory period in the transmission of nerve impulses. (2 marks)

Total: 5 marks

■ ■ ■

Candidates' answers to Question 2

Candidate A
(a) The increase in the membrane potential happens at the same time as the increase in sodium conductance. This is when the action potential occurs, so it must be due to the sodium ions. When the membrane potential falls back to normal again, this corresponds with an increase in potassium conductance.

A2 Biology

Candidate B
(a) The action potential is generated when the membrane potential becomes positive on the inside (it is usually −70 mV). The graph shows that this happens when the conductance to sodium ions increases, allowing these ions to rush in. To restore the membrane potential to −70 mV, potassium ions rush out. This can only happen when the membrane becomes permeable to potassium, which is shown by the increase in potassium ion conductance, after the peak occurs.

> Neither candidate has related the evidence to the terms 'depolarisation' and 'repolarisation'. Candidate A describes all the appropriate features in the graph, but has neither related them to the terms 'depolarisation' and 'repolarisation' nor related the evidence to the *events* of 'depolarisation' and 'repolarisation'. It is not at all clear that the candidate understands what these terms mean. Candidate B has not used the terms 'depolarisation' and 'repolarisation' but has described the *events* that are defined by the terms and has related the evidence to them. Candidate A scores 1 mark and Candidate B scores 2 marks. If you are asked to relate evidence to named events/processes, you will only score full marks if you show that you understand the required terms.

Candidate A
(b) The refractory period is the period when an axon cannot have an action potential. This is because it becomes permeable to sodium ions and can't let them pass through.

Candidate B
(b) This is the time when the membrane is impermeable to sodium and potassium ions, which means that a new action potential cannot be generated.

> Neither candidate really gets to grips with this. Both score just 1 mark for the idea that an action potential cannot be generated. Candidate A comes closest to the second mark, but makes contradictory statements by saying that the membrane 'becomes permeable to sodium ions' and 'can't let them pass through'. The examiner will not choose which of two contradictory statements a candidate really means: if the answer is not clear, it is wrong. Even without this, the answer is still not quite correct. During the refractory period it is not just that the membrane is impermeable (not strictly true, anyway) but also that it *cannot easily become permeable* to the ions. This is a subtle, but important, distinction.

> **This is a topic that many candidates find difficult. You must be quite clear about what graphs like this one show, such as the changes in polarity of the *inside* of the axon, compared with the *outside*. Also, you need to understand that making the *inside* more *positive* requires *positive* ions (sodium ions, Na⁺) to *enter*. Making the inside more negative again requires *positive* ions (potassium ions, K⁺) to *leave*. Notice that negative ions are not involved. Candidate A scores 2 marks, while Candidate B scores 3 marks.**

Question 3

Protein synthesis

Protein synthesis takes place in the ribosomes. The code for synthesis of a particular protein is specified by a section of the **DNA** molecule and is carried to the ribosomes by **mRNA**.

(a) (i) What do we call a section of DNA that codes for a protein? (1 mark)
    (ii) The **DNA** code is sometimes called a degenerate code. What does this mean? (2 marks)

(b) Figure 1 shows protein synthesis taking place in a ribosome.

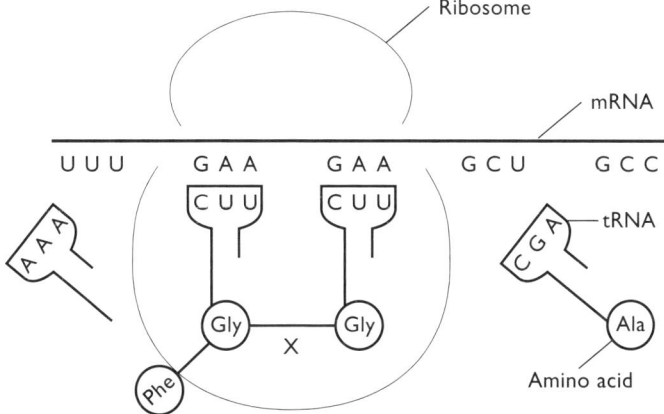

Key to amino acids
Phe = Phenylalanine
Gly = Glycine
Ala = Alanine

Figure 1

(i) Name the type of bond labelled **X**. (1 mark)
(ii) Use examples from the diagram to explain the terms *codon* and *anticodon*. (2 marks)

Total: 6 marks

Candidates' answers to Question 3

Candidate A
(a) (i) A gene

Candidate B
(a) (i) A small section of DNA that codes for a protein is called a gene.

📝 Both candidates know that a gene codes for a protein and each scores 1 mark.

Candidate A
(a) (ii) There are more codes than there are amino acids.

Candidate B
(a) (ii) Not all the triplet codes of the DNA code for amino acids; some are stop codes.

📝 Neither candidate answers the question completely. Each is awarded 1 mark for a correct statement. If they had each included the statement made by the other or had included the idea that some amino acids have more than one code, they would have scored both marks.

Candidate A
(b) (i) Glycosidic

Candidate B
(b) (i) Glycosidic bond

📝 Bonds between amino acids are *peptide* bonds. Both candidates appear to have been confused by the label 'gly' in the diagram. The key makes it quite clear that this is an abbreviation for the amino acid glycine. Look carefully at all the information given in a question.

Candidate A
(b) (ii) GAA is a codon; CUU is an anticodon

Candidate B
(b) (ii) A codon is a triplet of bases on the mRNA molecule — such as GAA. An anticodon is a triplet of bases on the tRNA molecule.

📝 Neither candidate really makes both points here. Candidate A scores 1 mark as (in this instance) the examples chosen can only be codon and anticodon respectively and also *they are complementary* — but is this good luck or does the candidate really understand that codon and anticodon must be complementary? Candidate B fails to score, despite probably understanding the concepts quite well. There is no example of an anticodon and there is no indication that codon and anticodon are complementary or that they code for an amino acid. The examiner will not assume *anything* for you: you must explain *everything*.

📝 **This is a topic that many candidates find difficult, as it pulls together knowledge from a number of areas. However, much of it depends only on learning the relevant facts. Candidate A scores 3 marks and Candidate B scores 2 marks.**

The polymerase chain reaction

The polymerase chain reaction (PCR) is a method of obtaining large amounts of DNA from a small initial sample. Figure 1 shows the main stages in the polymerase chain reaction.

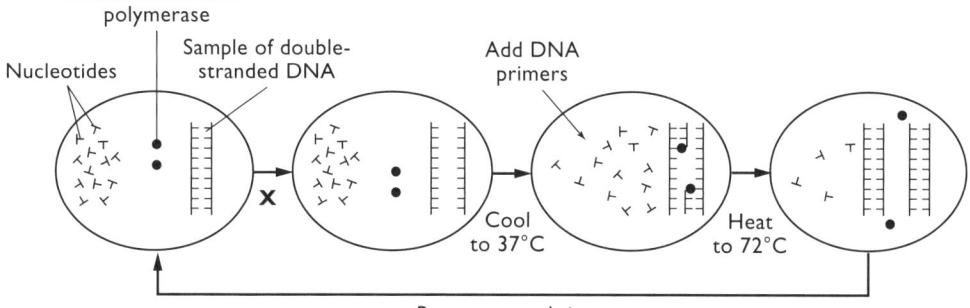

Figure 1

(a) (i) What must be done to separate the strands of DNA (process **X**)? (1 mark)
(ii) What are primers? (2 marks)

(b) (i) What is meant by 'thermostable DNA polymerase'? (2 marks)
(ii) What is the main advantage of using a thermostable DNA polymerase in this process? (1 mark)

Total: 6 marks

■ ■ ■

Candidates' answers to Question 4

Candidate A
(a) (i) You must heat them.

Candidate B
(a) (i) The DNA must be heated to 95°C to separate the strands.

> Both candidates score the mark here. There is no need to know the precise temperature — you are sitting a biology examination, not one in production engineering!

Candidate A
(a) (ii) Primers are small sections of DNA that start the new strands.

Candidate B

(a) (ii) Primers are small sections of single-stranded DNA that are complementary to one end of the original DNA strands.

> Candidate A clearly understands what primers do and, to an extent, what primers are. However, the key point that they are single-stranded is missing. Double-stranded DNA could not bind to the original strands in the way the primers do. Candidate A scores 1 mark. Candidate B knows that the primers are single-stranded and that their base sequences are complementary to those on the original strands. Although there is no mention of function, there is still enough information here to score 2 marks.

Candidate A

(b) (i) A thermostable DNA polymerase is one that is not easily denatured at high temperatures.

Candidate B

(b) (i) A thermostable DNA polymerase has a tertiary structure which is more stable and less easily deformed by heat.

> Both candidates explain 'thermostable' well — but neither describes what is meant by 'DNA polymerase', which is also included in the terms to be explained. Each scores 1 mark. Read the question carefully and make sure that you explain everything required. If a phrase is enclosed in quotes, make sure that you explain everything within the quotes.

Candidate A

(b) (ii) You can make the reaction faster by having it hotter.

Candidate B

(b) (ii) The enzyme will still be active at high temperatures. Probably its optimum temperature is high — it isn't easily denatured because of the strong bonds holding its tertiary structure.

> Candidate A explains the benefit — just, for 1 mark. A faster reaction means more product per minute for a manufacturer. Candidate B really only explains again (and in some detail) what 'thermostable' means, and fails to score. There is no description of an *advantage*.

> **Both candidates score 4 marks. This is an example of a question that is 'easy if you know the answers'. There are no difficult ideas and no complex data. So, did *you* know the answers?**

Gene cloning technology and its applications

Figure 1 shows how the action of enzyme **X** cuts a molecule of DNA.

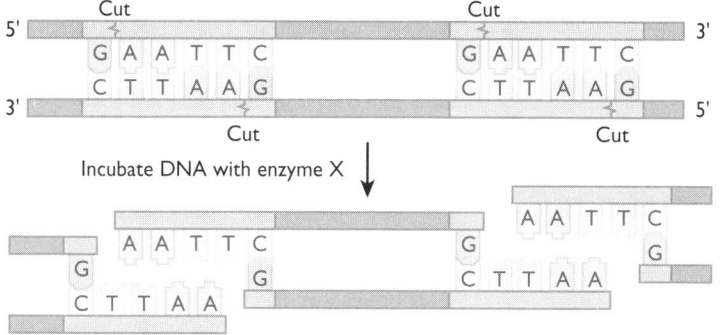

Figure 1

(a) (i)   Name the type of enzyme that cuts **DNA** in this way. (1 mark)
   (ii) What name is given to the places where the enzyme makes the cuts? (1 mark)
   (iii) Explain the importance of the type of cut made by this enzyme. (3 marks)

(b) Suggest why it may be preferable to obtain the gene from mRNA, rather than from **DNA**. (2 marks)

Total: 7 marks

■ ■ ■

Candidates' answers to Question 5

Candidate A
(a) (i) Restriction enzymes.

Candidate B
(a) (i) Restriction endonuclease.

   ✎ Each candidate scores 1 mark.

Candidate A
(a) (ii) Sticky ends.

Candidate B
(a) (ii) Restriction sites.

> Candidate A has confused the result of the cuts with the places in which the cuts are made. Candidate B is correct, for 1 mark.

Candidate A
(a) (iii) It makes a jagged cut so that the two ends will fit together again.

Candidate B
(a) (iii) The two ends that are produced have complementary shapes and so fit together again. They also have complementary base sequences.

> Neither candidate really explains the importance of restriction cuts, although both clearly understand that it relates to the future binding of DNA strands. Candidate B does make the point about complementary base sequences, but fails to follow that up by explaining:
> - why they are called 'sticky ends'
> - what they would be complementary with (other sections of DNA cut by the same enzyme)
> - the consequence of being complementary (hydrogen bonds could form between the sticky ends)
>
> Candidate A fails to score and Candidate B scores only 1 mark.

Candidate A
(b) This is because it will give a gene with no non-coding DNA.

Candidate B
(b) By using mRNA to produce the gene, it is likely that you will get more copies of the gene.

> The candidates each make one of the points in the mark scheme and each scores 1 mark.

> **Candidate A scores 2 marks and Candidate B scores 4. Both should have done better. This is a fairly uncomplicated question on genetic engineering. Most candidates who have prepared thoroughly would be expected to score 5 or 6 marks on such a question.**

Question 6

The control of gene action

Short interfering RNA (siRNA) is a type of RNA that is important in 'silencing' genes.

(a) Give *two* differences between siRNA and:
   (i) DNA (2 marks)
   (ii) mRNA (2 marks)

(b) Huntington's disease is a disorder in which the protein produced by a mutant gene causes progressive death of cells in the brain. The cells of sufferers from this condition frequently contain one mutant gene and one normal gene. Figure 1 shows how siRNA could be used in the treatment of such conditions.

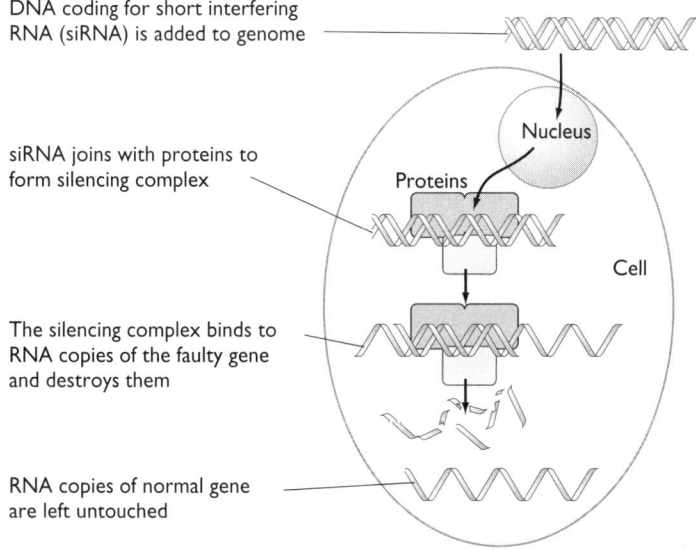

Figure 1

Use the diagram, and your knowledge of siRNA, to explain how siRNA might one day be used to treat Huntington's disease. (4 marks)

Total: 8 marks

■ ■ ■

Candidates' answers to Question 6

Candidate A
(a) (i) siRNA is smaller than DNA and it is single stranded.

Candidate B
(a) (i) DNA contains thymine but siRNA contains uracil. siRNA is single stranded.

> Both candidates make the mistake of assuming that siRNA is like other RNA and is single stranded. It is double stranded. Each candidate scores 1 mark only.

Candidate A
(a) (ii) It is smaller than mRNA.

Candidate B
(a) (ii) mRNA is a bigger molecule that has hydrogen bonds.

> Again, each candidates scores only 1 mark. They believe siRNA to be single stranded like other RNA and were, therefore, unable to offer its double stranded nature as a difference. Make sure of your facts. A single, simple misconception has cost these candidates 2 marks each — you cannot afford this. It is particularly disappointing in this question as the diagram for part (b) shows clearly that siRNA is double stranded. If you are unsure about an answer, look in other parts of the question for clues.

Candidate A
(b) DNA is introduced into the nucleus that codes for siRNA. The siRNA binds with proteins to form a silencing complex that binds to RNA copies of the faulty gene and destroys them.

Candidate B
(b) DNA that codes for siRNA is introduced and siRNA is produced that is specific to the mRNA made by the Huntington gene. When the siRNA joins with proteins to form a silencing complex, the complex joins to the Huntington mRNA and destroys it. This means that the Huntington mRNA cannot be translated into the Huntington protein and so this would reduce the symptoms of this terrible disease.

> Candidate A simple restates material that is given in the question. He/she does not even mention Huntington's disease, let alone explain how the treatment might work. The candidate fails to score. In a situation like this, where the question gives lots of relevant information, the expectation is that you will *apply* the information and make it relevant to the problem posed. Candidate B does this and makes at least four relevant points, for 4 marks.

> This is 'a question of two parts'. Part (a) demands recall of facts only and both candidates should have done better here. Part (b) is more demanding. However, with such questions make sure that you *use* the information supplied and do *not* merely restate it, as Candidate A did. **Candidate A scores only 2 marks, while Candidate B scores 6 marks.**

Auxins

The graph shows the effect of different concentrations of an auxin on the growth of stems and buds.

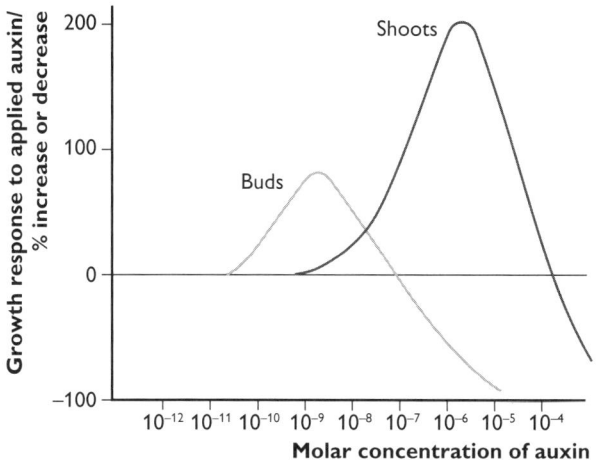

(a) What is auxin? (2 marks)

(b) Describe the effect of different concentrations of auxin on the growth of stems. (4 marks)

(c) Suggest why auxin inhibits the growth of side branches in growing stems. (3 marks)

Total: 9 marks

■ ■ ■

Candidates' answers to Question 7

Candidate A
(a) Auxin is a plant hormone that stimulates growth.

Candidate B
(a) Auxins are plant growth regulators. They can increase growth in stems and inhibit growth in roots.

   Candidate B gives a slightly better answer than Candidate A. However, neither candidate explains that auxins bring about their effects by influencing growth genes. Neither candidate really gives any more information than is to be found on the graph. Both candidates score 1 mark.

A2 Biology

Candidate A

(b) Went showed that increasing concentrations of auxin caused more growth, up to a maximum. Then there was no further increase in growth.

Candidate B

(b) As the concentration of auxin increases, the growth response increases. This happens until the concentration is 10^{-6} and then the growth response decreases.

> Candidate A does not relate the answer to the data supplied. However, there is a 'get-out-of-jail' card for Candidate A, because the question does not specifically ask candidates to use the data. Candidate A makes two valid points about Went's research that relate to the question and scores 2 marks. However, most data questions require you to use the data provided and, if you do not, you will fail to score.
>
> Candidate B does relate the answer to the data, but makes only two points, for 2 marks. Neither candidate has a chance of scoring 4 marks as each only makes two statements. To score all 4 marks the candidates should describe the trends (in the manner that Candidate B began to) by referring to the changes in growth response at specific concentrations. There is also the fact that for some concentrations the growth response is positive and for other (higher) concentrations, the growth response is negative.
>
> Look carefully at graphs and work out everything that they are telling you before you begin to describe the trends

Candidate A

(c) Shoots grow in quite high concentrations of auxin, but buds are inhibited by these concentrations.

Candidate B

(c) As the concentration of auxin increases, the growth response of buds decreases. Above 10^{-9} buds have a negative response. But stems are growing at these concentrations, so they will continue to grow but the side buds are stopped from growing.

> Both candidates show an understanding of the situation. Candidate A scores 2 marks and Candidate B scores 3 marks.
>
> **There are some fairly straightforward marks to be scored here of which the candidates have not taken full advantage. However, overall they have scored quite well considering that not all parts of the question are straightforward. Candidate A scores 5 marks and Candidate B scores 6 marks.**

Skeletal muscle

The sliding-filament hypothesis is believed to offer the best explanation of muscle contraction.

(a) The diagram shows part of a relaxed myofibril.

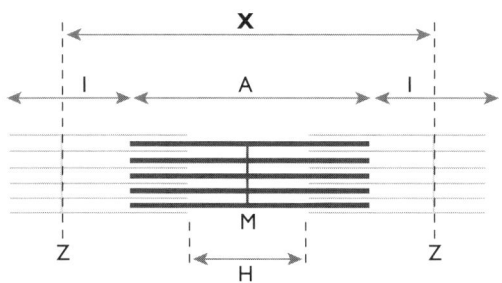

(i) Name the region labelled X. (1 mark)
(ii) Re-draw the diagram to show the appearance of the myofibril following contraction. (3 marks)

(b) Describe the roles of calcium ions and ATP in bringing about contraction of skeletal muscle. (4 marks)

(c) Give two differences between slow-twitch and fast-twitch fibres. (2 marks)

Total: 10 marks

■ ■ ■

Candidates' answers to Question 8

Candidate A
(a) (i) Muscle cell

Candidate B
(a) (i) Sarcomere

👉 Candidate B gives a more accurate answer than Candidate A and is awarded the mark. The region is not really a muscle cell.

Candidate A
(a) (ii)

Candidate B
(a) (ii)

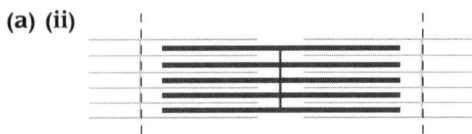

> Candidate A's diagram shows a reduction in the I band with the A band remaining unaltered. However, it is not clear whether or not there is an overall shortening of the sarcomere and the H band (which should be narrower) remains the same width. Candidate A scores 2 marks. Candidate B shows all the changes clearly, apart from the overall shortening of the sarcomere. Despite this, Candidate B scores the maximum 3 marks.

Candidate A
(b) Calcium ions open the binding sites on the actin and allow troponin to bind. ATP supplies the energy for the contraction.

Candidate B
(b) Calcium ions allow tropomyosin to bind with the actin. ATP releases energy to the myosin which causes the myosin to move and pull the actin filaments.

> Neither candidate really understands the role of calcium ions in allowing myosin to bind with actin, although Candidate A does mention that they 'open the binding sites'. However, this is the only mark Candidate A scores because the function of ATP is not described in sufficient detail. In the second sentence of his/her answer, Candidate B does rather better and scores 3 of the 4 available marks.

Candidate A
(c) Slow-twitch fibres contract more slowly than fast-twitch fibres and with less force.

Candidate B
(c) Slow-twitch fibres have more mitochondria and more myoglobin to supply oxygen.

> Candidate A probably thinks that his/her answer provides two valid points, but they are both combined in the same marking point. The candidate scores 1 mark. Candidate B scores 2 marks.

> **Candidate A scores 4 marks, while Candidate B scores 9. This is a question in which both knowledge and understanding of detail are important. There is no shortcut to that: you must learn the structure and function of skeletal muscle.**

Receptors and transmission of information through the nervous system

Sense cells, such as Pacinian corpuscles in the skin and rods and cones in the eye, are biological energy transducers. They transduce one form of energy from the environment into the electrochemical energy of a generator potential and an action potential. Nerve impulses may travel along neurones and across synapses to produce responses or they may be inhibited by action potentials from other neurones.

(a) Figure 1 shows the apparatus used in an investigation into the mode of action of Pacinian corpuscles. Figure 2 summarises the results obtained.

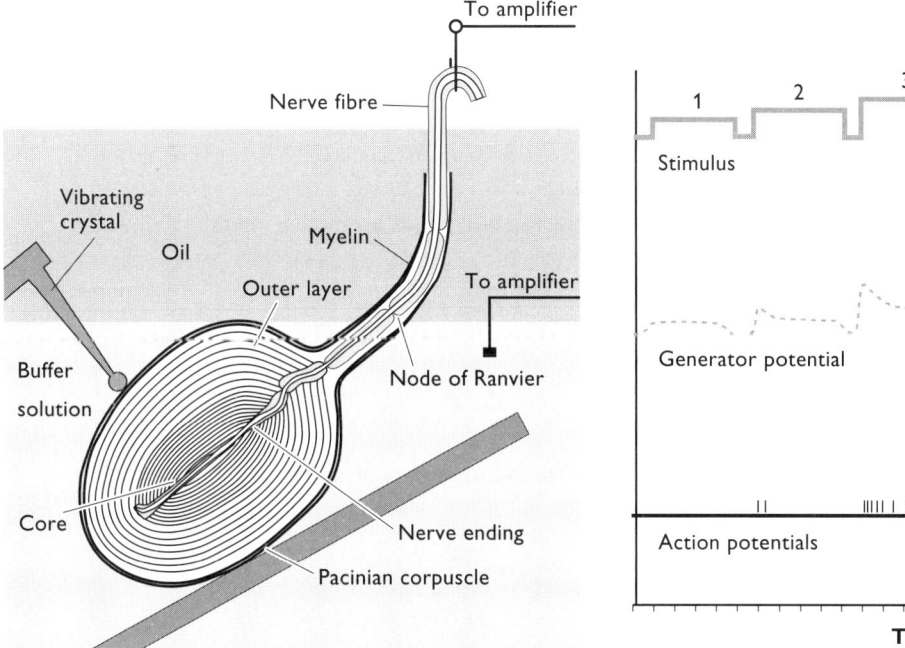

Figure 1   *Figure 2*

(i) How does the Pacinian corpuscle convert the vibrations of the crystal into a generator potential? (3 marks)

(ii) Suggest how the variations in the stimulus shown in Figure 2 could be generated using the above equipment. (2 marks)

A2 Biology

(iii) Use your knowledge of the nature of nerve impulses to explain the results obtained. (3 marks)

(b) Figure 3 and the table beneath it show how several sensory neurones can influence a single motor neurone.

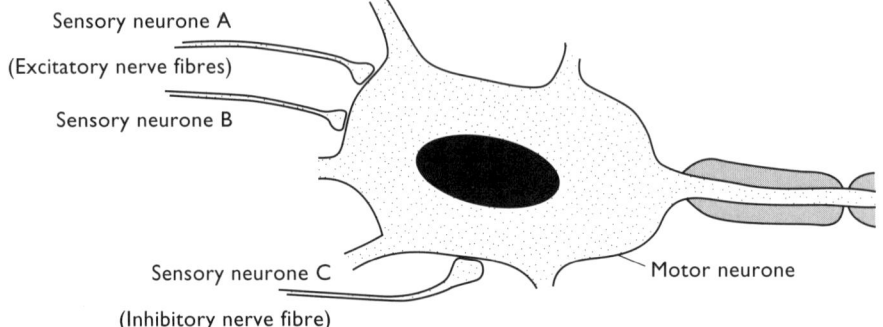

Neurone	Action potential				
Sensory neurone A (excitatory)	✗	✓	✗	✓	✓
Sensory neurone B (excitatory)	✗	✗	✓	✓	✓
Sensory neurone C (inhibitory)	✗	✗	✗	✗	✓
Motor neurone	✗	✗	✗	✓	✗

(i) Explain why a nerve impulse can only cross a synapse in one direction. (4 marks)
(ii) Explain the results shown in the table. (3 marks)

Total: 15 marks

■ ■ ■

Candidates' answers to Question 9

Candidate A
(a) (i) The vibrations of the crystal squash the Pacinian corpuscle and the pressure causes the nerve ending to let ions in and out.

Candidate B
(a) (i) The pressure of the crystal is transmitted through to the core of the Pacinian corpuscle where it affects the permeability of the membrane. Pressure-sensitive sodium ion channels are opened and sodium ions enter to create a generator potential.

> Candidate A understands the basic ideas about how a Pacinian corpuscle functions, yet fails to score any marks. The answer is simply too vague. Compare

this with the same ideas expressed in much more detail and with no ambiguity in Candidate B's answer, which scores all 3 marks. You must revise in detail.

Candidate A

(a) (ii) The crystal vibrates and the oil absorbs some of the energy of the vibration. If you used less oil, the vibrations would be bigger.

Candidate B

(a) (ii) One could apply a larger voltage to make the crystal vibrate more.

> Both candidates make sensible suggestions, but Candidate A's answer is *explained* in more detail and earns both marks. Candidate B scores just 1 mark.

Candidate A

(a) (iii) When the stimulus gets larger, the generator potential also gets bigger and you get more action potentials. This is because the action potential can't get bigger — it's all-or-nothing.

Candidate B

(a) (iii) With more stimulation, the increased pressure opens more sodium ion channels and the generator potential is increased. Because the generator potential generates the action potential, there ought to be a bigger action potential as well. But action potentials are 'all-or-nothing' and so, instead of getting a bigger action potential, you get more action potentials.

> Candidate B demonstrates a clear understanding of the situation and scores all 3 marks. Candidate A *describes* the relationship between stimulus, generator potential and action potential, but only *explains* why an increase in the generator potential does not produce an increase in the action potential, and so scores only 1 mark. Make sure you *explain* what you are asked to explain.

Candidate A

(b) (i) The ends of the pre-synaptic neurones contain vesicles of neurotransmitter. When an action potential reaches the end, these vesicles release their neurotransmitter and it diffuses across to the post-synaptic membrane. It can only pass one way because of the concentration gradient.

Candidate B

(b) (i) The synaptic knob of the pre-synaptic neurone contains vesicles that produce the neurotransmitter. When an action potential arrives at the synaptic knob, calcium ions enter and cause the vesicles to move to the surface and release their neurotransmitter into the synaptic cleft. It crosses the synaptic cleft and binds to protein receptor molecules on the post-synaptic neurone. This causes an action potential to be generated here and the nerve impulse is transmitted along the second neurone.

> Candidate A scores 2 of the 4 marks, for explaining how neurotransmitters cross the synaptic cleft (by diffusion) and why they pass in the direction they do

A2 Biology

(because of the concentration gradient). Candidate B gives a full account of how synaptic transmission takes place, but does not answer the question asked and so does not score any marks. The issue is why the neurotransmitter *only* crosses the synapse in *one direction*. To score full marks, four of the following points must be made:

- Only the pre-synaptic neurone has vesicles that produce the neurotransmitter.
- Neurotransmitter diffuses across the synaptic cleft from the pre-synaptic membrane to the post-synaptic membrane.
- It diffuses in this direction because of the concentration gradient (higher near the pre-synaptic neurone).
- Binding to the receptor protein on the post-synaptic membrane removes transmitter from the cleft and maintains the gradient.
- The neurotransmitter is hydrolysed before being released by the post-synaptic membrane back into the synaptic cleft.

All the information about action potentials arriving and calcium ions entering is not relevant to this question. Make sure you read the question carefully and understand what is being asked. Don't go into unnecessary detail or get sidetracked into irrelevancies.

Candidate A

(b) (ii) When several neurones act independently like this on another neurone, the effect is called summation. You only get an action potential in the motor neurone when A and B act together. If A and B act alone, or with C, there is no action potential in the motor neurone.

Candidate B

(b) (ii) Neurones A or B do not cause an action potential in the motor neurone on their own because they do not release enough neurotransmitter and so the threshold is not reached. When both have an impulse and both release neurotransmitter, their combined effect is enough to pass the threshold and start an action potential. When either A or B or both A and B have an impulse with C, the inhibitory effect of C is enough to counteract A and B — together or separately. This is called summation.

> Neither candidate *explains* summation satisfactorily. Candidate A suggests that it occurs when neurones act *independently* — probably a slip, as in the next line the neurones are described as acting together. The rest of Candidate A's answer is a *description* of the results, not an *explanation*, and so the candidate fails to score. Candidate B explains why A and B separately do not produce action potentials and why, together, they do. However, the answer does not *explain* the inhibitory effect of neurone C. Candidate B scores 2 marks. Be clear about what the question is asking. An examiner would look for the following points:
> - Neurones A and B acting independently produce sub-threshold stimulation.
> - Together, A and B exceed the threshold.

- This is called summation.
- Neurone C can prevent A and/or B exceeding the threshold.

☑ **The question demands a detailed understanding of Pacinian corpuscles and how transmission of nerve impulses takes place. You need to revise the material carefully and in detail. Candidate A could quite easily have scored 4 or 5 more marks here if more detail had been supplied. The answers appear to show** *understanding*, **but there is a lack of detail. There is no short cut to this and, equally, no magic — you must simply put the time in. Candidate A scores 5 marks and Candidate B scores 9.**

Homeostasis

Within set limits, humans are able to maintain a plasma glucose concentration and high body temperature that vary only slightly.

(a) (i) Explain the role of pancreatic hormones in maintaining the plasma glucose concentration within set limits. (6 marks)

(ii) Explain why these pancreatic hormones are able to target liver and skeletal muscle cells in particular. (4 marks)

(b) Explain the benefit of being able to maintain a constant, high body temperature. (5 marks)

Total: 15 marks

■ ■ ■

Candidates' answers to Question 10

Candidate A
(a) (i) When there is too much glucose in the blood, the pancreas produces insulin to bring the level down. When there isn't enough glucose in the blood, the pancreas produces glycogen to raise the level. These two hormones maintain the level of glucose in the blood within set limits.

Candidate B
(a) (i) If the plasma glucose concentration exceeds set limits, β-cells in the islets of Langerhans secrete insulin. Insulin stimulates liver cells to absorb glucose and convert it to glycogen. Glucagon, secreted by α-cells, has the opposite effect.

Candidate A provides a simplistic account of the action of the pancreatic hormones *and* confuses glucagon with glycogen. You *must* be able to distinguish between the two. Candidate B describes the role of insulin accurately. However, this is only one half of the story. Writing that 'glucagon does the opposite' is not enough. The examiner will not re-interpret the points you have made. In a situation like this you must describe the role of both hormones. Candidate A scores 1 mark while Candidate B scores 3 marks. If Candidate B had described the circumstances under which glucagon is secreted, another mark would have been allocated.

Candidate A
(a) (ii) Insulin can target liver and muscle cells because there are receptors on these cells, but there aren't receptors on other cells.

Candidate B
(a) (ii) Receptors on the surface of liver and muscle cells have a specific shape to which molecules of insulin and glucagon can bind. These receptors are not present on other cells.

> Neither candidate gives a full account. Candidate B gives a little more information than Candidate A and scores 2 marks; Candidate A scores just 1 mark. Neither candidate describes the nature of these receptors or mentions the complementary shape of the hormone molecules.

Candidate A

(b) Maintaining a constant, high body temperature means that you don't depend on the environment to warm you up or cool you down. As the temperature is always the same, all the processes in the body can carry on at the same rate because all biological processes are affected by temperature.

Candidate B

(b) A constant, high body temperature means that humans are independent of their environment and can colonise inhospitable areas of the planet. A body temperature of 37°C means that the enzymes of our body are always working at their optimum and so all the processes are carried out quickly. If our body temperature were higher than this, the enzymes would be denatured.

> Both candidates appreciate that being able to maintain a constant body temperature gives humans a degree of independence from their environment. Candidate A appreciates that temperature affects the rate of reactions, but does not mention that enzymes control the reactions. Candidate B appreciates that enzymes are involved, but does not explain adequately how temperature affects reactions. Candidate A scores 2 marks and Candidate B scores 3.
>
> To score full marks on this question, an examiner would expect five of the following points:
> - All biological processes are influenced by temperature.
> - Increasing the temperature increases the rate of biological processes.
> - Biological processes are controlled by enzymes.
> - Too high a temperature would denature the enzymes.
> - A variable temperature would mean that the enzymes would not always be working at their optimum.
> - A constant high temperature gives an organism independence from its environment.

> Homeostasis is a topic that most candidates think they understand. Often they do understand the main principles but, as shown in the example above, marks are frequently lost by not supplying sufficient detail. Candidate A scores just 4 out of 15 marks and Candidate B scores only 8.

> **Throughout the questions, Candidate B performs better than Candidate A. Overall, Candidate B scores 56 marks, whereas Candidate A scores just 32 out of a possible 86 marks. This probably represents a grade-B performance for Candidate B and a grade-E performance (just) for Candidate A.**

This may be in part due to ability and preparation, but there are a number of self-inflicted injuries in Candidate A's performance. On several occasions, Candidate A has clearly understood the biology required in the answers, but has not supplied the detail. You cannot hope to be awarded 5 marks for writing a couple of lines. You must also make sure that, when supplied with information in a question, you use the information and do not just re-state it.

Make sure that you:
- read the question carefully
- take note of the verb used (e.g. are you being asked to describe, explain, analyse?)
- take careful note of the subject of the question and of any qualification (e.g. for how long, between certain times, over region A of the graph)
- use data from a diagram/table/graph in your answer when specifically instructed to do so

Make sure that you tailor your answers to the above criteria and to the mark allocation.